Sociotechnical Systems: A Sourcebook

Sociotechnical Systems: A Sourcebook

Edited by

William A. Pasmore

and

John J. Sherwood

University Associates, Inc.
7596 Eads Avenue
La Jolla, California 92037

ISBN: 0-88390-142-0
Library of Congress Catalog Card Number 77-20543
Printed in the United States of America

To Bonnie and Florence

Contents

Foreword

Writing a foreword is a narcissistic experience for the individual asked to assume this task, since it is often assumed that that person is unusually knowledgeable. When Professors Pasmore and Sherwood asked me to write this piece, I was simultaneously flattered and frightened: flattered for the recognition and frightened about my lack of accumulated wisdom. The field of sociotechnical systems has been adorned by such outstanding innovators as A. K. Rice, Fred Emery, Eric Trist, Eric Miller, Louis Davis, Richard Walton, Wilfred Brown, James G. Miller, and younger colleagues who have followed in their footsteps throughout the world. But for me to write a statement regarding these contributions is sheer narcissistic fantasy at best and a misrepresentation of the issues at worst. Nevertheless, the task is exciting and challenging. It has caused me to think and brood for a long time. Those who read forewords in a significant volume of this kind are undoubtedly true believers, and their attention deserves special treatment. I shall not attempt to enumerate the many contributions of the various approaches to the field of sociotechnical systems, but I will issue a mandate for some broad questions and new discoveries in the applied behavioral science field as a whole.

It is a matter of great satisfaction to me that the applied behavioral sciences have been enriched both in theory and in practice by the contributions of those in the field of sociotechnical systems and by experiments to improve the quality of work life. Lewin's dictum that "the most practical thing in the world is a good theory" has been practiced in its fullest sense by sociotechnical system innovations. This volume is a witness to the fact that the applied behavioral sciences have made significant progress in developing organizational change strategies and that experiments in sociotechnical systems have provided a particularly viable agenda for the future of planned organizational change methodology. It is a non sequitor to write a foreword for a volume in a field that is so full of success stories and that continues to generate enthusiasm among both academicians and management practitioners.

However, no work in the applied behavioral sciences can ever be completely comprehensive. Therefore, I have chosen to highlight two sides of the field: one, the fine points that have been basically responsible for the success of the sociotechnical system approach and, two, some unanswered questions that remain on the agenda for applied behavioral science researchers.

The good points are easier to articulate. The potency and strength of the sociotechnical system concept lies in its experimental approach to planned organizational change. The contents of this volume are solid testimony that the ideas generated by sociotechnical system experimenters continue to be robust and challenging—robust because they are rooted in continuing efforts to refine variables and rearrange experimental designs and challenging because they use knowledge from various disciplines that bear on the issues of the quality of work life.

A close look at the work in the field, especially the articles in this volume, indicates that sociotechnical system research continues to flourish because of its contributions to planned organizational change. Earlier assumptions of planned organizational change concepts were quite similar in logic to those subscribed to by the sociotechnical system approaches. The most fundamental assumption operative in both designs is rationality. Both theories are goal directed and improvement oriented and have a futuristic reference. Both rely heavily on logical and conceptual forms of description, prediction, control, and change embedded in the simple definition of experimental methodological processes.

Another contribution of the sociotechnical system concept has been the benign and pragmatic technological ease that makes the concept manageable for experimental purposes. It has belied the complexity of human experience and caught the managerial imagination for introducing technological transformation for the betterment of organizational goals. Arranging people around a well-defined, structured technology was far easier than arranging a complex set of human relationships in an organizational context. As long as technology determined the limits of work structure, it was possible and feasible to organize the work force around the system. Because of this, sociotechnical system approaches have had a heyday in traditional technologies such as textiles, coal mining, and even mass assembly lines.

But what have been the artifacts of sociotechnical experiments? What philosophical and practical questions remain unanswered? What contributions are possible in the future? Although complete answers cannot be attempted in a foreword, I wish to articulate a few issues for future consideration.

First, researchers must address themselves to some of the managerial dilemmas of ends vs. means, a mechanistic vs. a humanistic outlook, personal vs. sociocentric styles, and the efficiency vs. the effectiveness of an organization. The haunting questions of managers remain the same in any planned organizational change methodology: Can we continue to organize work systems simply to increase productivity? Do increased productivity and efficiency contribute to the quality of work life? Are efficient methods the same as effective methods? Can we organize the blue-collar workers and leave the managerial cadre untouched and independent of our experiments? Can we continue to utilize entrepreneurial and managerial skills within the same theoretical framework? Can we continue to assume the premise of rationality in management thinking, even when evidence is mounting against it? Of course, similar questions can be asked of other planned organizational change strategies, but the onus is on us as sociotechnical system designers to respond with concepts, data, values, and new experimental designs.

Second, it is my contention that the theories and experiments in the field of sociotechnical systems have not been directed toward the day-to-day realities of organizational life. The hierarchical nature of organizational decision making, the prevalent political processes in organizations, the reality that machines and technology come directly in contact only with workers and not with managers are some of the major facts confronting the organizational change planner. For these reasons, the long-term consequences of sociotechnical system interventions are not known, and the challenge continues to be to provide this data.

Third, amazing as it seems, ideas and experiments in sociotechnical systems continue to be localized. A great dearth of such experiments exists in the United States. The earlier experiments were done in coal mines in the United Kingdom and in textile mills in India. Later these experiments were substantiated by European scholars through their conceptualizations and experiments. In the United States, most of the work that has been done has melded the issues of the quality of work life with the experiments in sociotechnical systems. Does this situation mean that technologically advanced countries find it difficult to experiment with new sociotechnical systems? Or have applied behavioral scientists in the United States been able to use a number of other meaningful intervention strategies with equal success? The answers to these questions remain a matter of goodwill and gesture, until we do some thorough analyses of the issues; I hope that additional research will be one consequence of this volume.

Notwithstanding the unsolved issues, the contributions of sociotechnical system concepts and experiments continue to be a major thrust in the field of organizational change. I wish to commend my colleagues, Professors Pasmore and Sherwood, for bringing together this significant volume, which will be a great resource for future thought and experimentation.

Case Western Reserve University *Suresh Srivastva*
December 1977

PART I:

ORGANIZATIONS AS SOCIOTECHNICAL SYSTEMS

Organizations as Sociotechnical Systems

William A. Pasmore and John J. Sherwood

Sociotechnical system interventions are organization development techniques that typically involve the restructuring of work methods, rearrangements of technology, or the redesign of organizational social structures. The objective is to optimize the relationship between the social or human systems of the organization and the technology used by the organization to produce output. When these systems are arranged optimally, the organization runs more smoothly than when they are not; output is higher, employees' needs are satisfied better, and the organization remains adaptable to change.

In practice, working toward the joint optimization of the social and technological systems of an organization is a complex process that requires a thorough understanding of: (1) the social processes that occur in organizations and the variety of theories and methods that exist to make more efficient use of human resources; (2) the technological process used by the organization and the constraints that it places on the design and operation of the social system; (3) the theory of open systems, because no two organizations are exactly alike or are faced with the same environmental demands; and (4) the mechanics of change, both in the execution of the initial sociotechnical system design and in provision for the continual adaptation of the organization to new environmental demands.

Each sociotechnically designed organization is unique because of internal and external environmental differences, so no single method of intervention can or should be advocated. Resource articles are included in this text that provide an exposure to many different methods, leaving the decision to the reader about which is useful in a given situation.

THE ORGANIZATION AS AN OPEN SYSTEM

Sociotechnical system theorists view an organization as an open and living system, much like a biological cell, i.e., engaged in active transactions with the environment. Raw materials are imported, and finished goods or services are exported. The environment, through legislation, public demand, or the competitive actions of other organizations, may also exert pressures on the organization to follow certain rules or to structure itself in a given way. Products become obsolete, new legislation is passed, the economy changes, populations shift in communities, and social norms are altered. These and many other factors have an impact on the internal design and functioning of an organization. It is important, therefore, that the organization be aware of environmental changes when optimizing its social and technical systems.

Some organizations, e.g., hospitals, have rapidly changing environments; others, e.g., grocery stores, have relatively stable environments over a period of time. Even if their environments are stable, all organizations must learn to monitor and meet the demands of that environment. Otherwise, organizations always face the possibility of extinction due to obsolescence of product or internal inefficiency.

SOCIOTECHNICAL SYSTEM THEORY

Many theories of organizational behavior view an organization's technology as fixed or given and, therefore, not amenable to change. If organizational improvement is the goal, then making more efficient use of human resources may not be sufficient; changes in technology may be needed to affect performance in any major or enduring way. Although the environment cannot be ignored, the relationship between the social and technical systems within the firm also must be taken into account.

Sociotechnical system theory provides guidance and direction for the exploration of an organization and goals for change. It describes the complex relationships between people, tasks, and technologies and helps us to see how these can be used to advantage. One position of sociotechnical theorists, for example, is the need to design jobs that give employees control over the key variances of an organization's production or service process. This has typically been accomplished by forming self-directing autonomous work groups that have major responsibility for task performance.

Sociotechnical theory has evolved into a set of fairly stable and recognizable propositions. These specify (1) that the design of the organization must fit its goals; (2) that employees must be actively involved in designing the structure of the organization; (3) that variances in production or service must be controlled as close to their source as possible; (4) that subsystems must be designed around relatively whole and recognizable tasks; (5) that support systems must be congruent with the design of the organization; (6) that a high quality of work life should be provided; and (7) that changes should continue to be made as necessary to meet environmental demands.

Although sociotechnical theory is itself susceptible to change and refinement, it does provide the concepts needed to facilitate organizational diagnosis and change efforts.

SOCIOTECHNICAL DIAGNOSIS AND CHANGE

Open system theory and sociotechnical system theory give the manager, consultant, or practitioner background for undertaking organizational diagnosis and change. Facilitating such change is easier because so many studies have been done. Data about an organization's technology, structure, personnel, and environment can be collected and used.

Typically, the diagnostic process is done in a relatively fixed order. The change agent collects, in order, (1) general information about the organization, which orients the analyst to the specific setting; (2) specific information about the production or service process, with special attention to the variances that occur and how and by whom they are controlled; (3) data about the quality of working life to identify needs to be met; and (4) data on the organizational environment and any ongoing organizational change programs. This information provides indications of areas in which changes can be made to more nearly meet *both* the goals of the organization and the needs of the individuals working within it.

Once the analyst has performed an adequate diagnosis, he or she has the task of inducing change in the organization. This is often difficult, and it may at times seem more an art than a science. The change agent must know something about the dynamics of change processes and problems such as resistance and diffusion that may occur.

Because every organization is unique, sociotechnical diagnosis reveals different targets for change in different settings. The change agent must share knowledge with members of the organization; the purpose and direction of the change process must be made clear to them; and those who must implement the change and work within the new organizational structure must accept it.

Exactly how the change process should be undertaken to ensure optimal results is not very clear. At this stage, it is probably best to study the models presented by experts, to read reports of actual interventions, and to develop an internalized model that is acceptable. We hope that the answer to the question "What do we really know about sociotechnical systems?" will become clear.

SOCIOTECHNICAL STUDIES: WHAT DO WE REALLY KNOW?

From time to time, it is wise to pause, consider where we have been, and make suggestions about the directions that further research must take. Assessing the state of the art keeps us from becoming overconfident; it also inspires further research to discover new and reshape current knowledge.

What we *do* know about sociotechnical system theory and intervention has filled a great number of books and journals; what we still *want* to know could easily fill the same space and much more. We have learned that sociotechnical methods seem effective, given the right circumstances and the proper application of the methods. Furthermore, we have come to believe that sociotechnical techniques are more effective and longer lasting for improving organizational performance than most other applied behavioral science interventions.

What we do not know yet are just which conditions and techniques are most likely to ensure successful change. Also in need of much refinement are our definitions of technology, of human needs, and of the jointly optimized relationships between them in various settings.

The solutions to these and other puzzles are being uncovered slowly. Studies are reviewed, comparisons are made, and knowledge is advanced. As the process continues, new ideas and challenges are presented so that we know more about what things we do know and what things we do not know. It seems clear that sociotechnical system interventions *can* work; therefore, they will continue to be the focus of much interest and effort on the part of managers, practitioners, and researchers.

PERSPECTIVES ON SOCIOTECHNICAL SYSTEMS

At various points in the evolution of any theory, experts in the field provide their perspectives on the theory's development and the extent of its application. In the case of sociotechnical theory, most experts seem convinced of the value of the theory, although all have some reservations about its future and have their own ideas about the directions that future research should take.

Our own feeling is that sociotechnical system theory has been of great benefit to the cultures in which it has been applied—primarily blue-collar, production-oriented organizations. However, we believe strongly that two challenges must be met in the future: (1) sociotechnical system theory must be expanded and redefined to deal with the social and task systems of all organizations, including those that do not utilize well-defined mechanical technologies, such as hospitals, universities, and the administrative staffs of most firms; (2) the supervisor must play an integral role in the design and implementation of sociotechnical systems. Both recommendations underscore the need to expand our understanding of the social dynamics of organizational change for the benefit of those involved.

Thus, although the study of sociotechnical systems has become and should continue to be a solidly entrenched discipline in the applied behavioral sciences, much of its potential is yet to be discovered and utilized. This remains an important task for the future.

PART II:

WORKING WITH THE ORGANIZATION AS AN OPEN SYSTEM

Introduction

Because organizations do not exist in vacuums, when we speak of jointly optimizing the relationship between social and technological systems, we also must consider explicitly the impact of the firm's environment, which plays a vital role both in shaping organizations and in the design of sociotechnical systems. From the environment the organization obtains raw materials, labor, capital, and expertise; the demand for organizations, products, or services also comes from the environment. In this sense, organizations are considered "open" to their environments, and we can speak of designing "open" sociotechnical systems.

When groups exist for the satisfaction of their members alone, they are not truly organizations by our definition because they do not produce an output or satisfy a need arising from their environment. Street gangs, coffee klatsches, or bridge groups, because they do not exist to satisfy environmental demands, are referred to here as relatively "closed" systems. Although most of this volume is concerned with open systems, closed systems are discussed from time to time because they do provide comparative data.

More than ever, managers today must be concerned with the impact of the environment on their organizations and, of course, the impact of their organizations on the environment. Besides the procurement of resources from and the dispersal of finished goods and services to the environment, an organization must be concerned with a third major environmental force: government(s). The growing size of corporate legal staffs, new specialists devoted to meeting guidelines dictated by privacy, the Occupational Safety and Health Act, and Equal Employment Opportunity, and the growing cost of malpractice insurance indicate the increasing role of government in shaping the internal life of organizations. Various governmental agencies no longer have secondary or marginal influence. Organizations must be responsive to legislative changes, if not proactive in shaping their development. Supply, demand, and legislation provide limits within which sociotechnical systems are designed. For this reason, we believe it is fitting to begin this collection with articles that explicate the transactions that occur between organizations and their environments.

Emery and Trist, in their classic work, "The Causal Texture of Organizational Environments" deal primarily with changes that are constantly occurring within the environment. These changes are not promulgated by the actions of an organization, but the organization is often greatly influenced by them. For example, the changing tastes of consumers or the actions of competitors often affect the product strategy of a manufacturer, although both are largely beyond the manufacturer's control. Emery and Trist define four types of causal textures, ranging from the placid, randomized environment of the economist's model of perfect competition to the highly complex turbulence that seems characteristic of much of the contemporary world.

As environments become more complex, organizations are faced with greater uncertainty; according to Emery and Trist, organizations facing turbulent environments require forms of organizational structure that are substantially different from the traditional hierarchy to which we are accustomed.

For the reader interested more in action than in theory, Jayaram's piece is a guide to "Open Systems Planning." Jayaram designs a procedure to give organizational members an opportunity to identify rapidly and systematically the relevant aspects of their environments and then to make deliberate plans for responding reactively or proactively to new trends. According to Jayaram, because the environment has a critical impact on organizational design, it is desirable for open system planning to be the precursor—the first concrete step—to any sociotechnical system intervention.

These two articles set theoretical and practical guidelines for understanding and applying the concepts and techniques that follow. They also suggest possible directions for further research that is vital to further developments in sociotechnical systems.

The Causal Texture of Organizational Environments

F. E. Emery and E. L. Trist

IDENTIFICATION OF THE PROBLEM

A main problem in the study of organizational change is that the environmental contexts in which organizations exist are themselves changing, at an increasing rate, and towards increasing complexity. This point, in itself, scarcely needs labouring. Nevertheless, the characteristics of organizational environments demand consideration for their own sake, if there is to be an advancement of understanding in the behavioural sciences of a great deal that is taking place under the impact of technological change, especially at the present time. This paper is offered as a brief attempt to open up some of the problems, and stems from a belief that progress will be quicker if a certain extension can be made to current thinking about systems.

In a general way it may be said that to think in terms of systems seems the most appropriate conceptual response so far available when the phenomena under study—at any level and in any domain—display the character of being organized, and when understanding the nature of the interdependencies constitutes the research task. In the behavioural sciences, the first steps in building a systems theory were taken in connection with the analysis of internal processes in organisms, or organizations, when the parts had to be related to the whole. Examples include the organismic biology of Jennings, Cannon, and Henderson;

Reprinted from F. E. Emery & E. L. Trist, "The Causal Texture of Organizational Environments," HUMAN RELATIONS, 1965, *18*, 21-31. Used with permission of Plenum Publishing Corporation.

early Gestalt theory and its later derivatives such as balance theory; and the classical theories of social structure. Many of these problems could be represented in closed-system models. The next steps were taken when wholes had to be related to their environments. This led to open-system models.

A great deal of the thinking here has been influenced by cybernetics and information theory, though this has been used as much to extend the scope of closed-system as to improve the sophistication of open-system formulations. It was von Bertalanffy (1950) who, in terms of the general transport equation which he introduced, first fully disclosed the importance of openness or closedness to the environment as a means of distinguishing living organisms from inanimate objects. In contradistinction to physical objects, any living entity survives by importing into itself certain types of material from its environment, transforming these in accordance with its own system characteristics, and exporting other types back into the environment. By this process the organism obtains the additional energy that renders it 'negentropic'; it becomes capable of attaining stability in a time-independent steady state—a necessary condition of adaptability to environmental variance.

Such steady states are very different affairs from the equilibrium states described in classical physics, which have far too often been taken as models for representing biological and social transactions. Equilibrium states follow the second law of thermodynamics, so that no work can be done when equilibrium is reached, whereas the openness to the environment of a steady state maintains the capacity of the organism for work, without which adaptability, and hence survival, would be impossible.

Many corollaries follow as regards the properties of open systems, such as equifinality, growth through internal elaboration, self-regulation, constancy of direction with change of position, etc.—and by no means all of these have yet been worked out. But though von Bertalanffy's formulation enables exchange processes between the organism, or organization, and elements in its environment to be dealt with in a new perspective, it does not deal at all with those processes in the environment itself which are among the determining conditions of the exchanges. To analyse these an additional concept is needed—*the causal texture of the environment*—if we may re-introduce, at a social level of analysis, a term suggested by Tolman and Brunswik (1935) and drawn from S. C. Pepper (1934).

With this addition, we may now state the following general proposition: that a comprehensive understanding of organizational behaviour requires some knowledge of each member of the following set, where L indicates some potentially lawful connection, and the suffix 1 refers to the organization and the suffix 2 to the environment:

$$L_{11}, L_{12}$$
$$L_{21}, L_{22}$$

L_{11} here refers to processes within the organization—the area of interdependencies; L_{12} and L_{21} to exchanges between the organization and its environment—the area of transactional interdependencies, from either direction; and L_{22} to processes through which parts of the environment become related to each other—i.e. its causal texture—the area of interdependencies that belong within the environment itself.

In considering environmental interdependencies, the first point to which we wish to draw attention is that the laws connecting parts of the environment to each other are often incommensurate with those connecting parts of the organization to each other, or even with those which govern the exchanges. It is not possible, for example, always to reduce organization-environment relations to the form of 'being included in'; boundaries are also 'break' points. As Barker and Wright (1949), following Lewin (1936), have pointed out in their analysis of this problem as it affects psychological ecology, we may lawfully connect the actions of a javelin thrower in sighting and throwing his weapon; but we cannot describe in the same concepts the course of the javelin as this is affected by variables lawfully linked by meteorological and other systems.

THE DEVELOPMENT OF ENVIRONMENTAL CONNECTEDNESS (CASE I)

A case history, taken from the industrial field, may serve to illustrate what is meant by the environment becoming organized at the social level. It will show how a greater degree of system-connectedness, of crucial relevance to the organization, may develop in the environment, which is yet not directly a function either of the organization's own characteristics or of its immediate relations. Both of these, of course, once again become crucial when the response of the organization to what has been happening is considered.

The company concerned was the foremost in its particular market in the food-canning industry in the U.K. and belonged to a large parent group. Its main product—a canned vegetable—had some 65 percent of this market, a situation which had been relatively stable since before the war. Believing it would continue to hold this position, the company persuaded the group board to invest several million pounds sterling in erecting a new, automated factory, which, however, based its economies on an inbuilt rigidity—it was set up exclusively for the long runs expected from the traditional market.

The character of the environment, however, began to change while the factory was being built. A number of small canning firms appeared, not dealing with this product nor indeed with others in the company's range, but with imported fruits. These firms arose because the last of the post-war controls had been removed from steel strip and tin, and cheaper cans could now be obtained in any numbers—while at the same time a larger market was developing in imported fruits. This trade being seasonal, the firms were anxious to find a way of using their machinery and retaining their labour in winter. They became able to do so through a curious side-effect of the development of quick-frozen foods, when the company's staple was produced by others in this form. The quick-freezing process demanded great constancy at the growing end. It was not possible to control this beyond a certain point, so that quite large crops unsuitable for quick freezing but suitable for canning became available—originally from another country (the United States) where a large market for quick-frozen foods had been established. These surplus crops had been sold at a very low price for animal feed. They were now imported by the small canners—at a better but still comparatively low price, and additional cheap supplies soon began to be procurable from underdeveloped countries.

Before the introduction of the quick-freezing form, the company's own canned product—whose raw material had been specially grown at additional cost—had been the premier brand, superior to other varieties and charged at a higher price. But its position in the product spectrum now changed. With the increasing affluence of the society, more people were able to afford the quick-frozen form. Moreover, there was competition from a great many other vegetable products which could substitute for the staple, and people preferred this greater variety. The advantage of being the premier line among canned forms diminished, and demand increased both for the not-so-expensive varieties among them and for the quick-frozen forms. At the same time, major changes were taking place in retailing; supermarkets were developing, and more and more large grocery chains were coming into existence. These establishments wanted to sell certain types of goods under their own house names, and began to place bulk orders with the small canners for their own varieties of the company's staple that fell within this class. As the small canners provided an extremely cheap article (having no marketing expenses and a cheaper raw material), they could undercut the manufacturers' branded product, and within three years they captured over 50 percent of the market. Previously, retailers' varieties had accounted for less than 1 percent.

The new automatic factory could not be adapted to the new situation until alternative products with a big sales volume could be developed, and the scale of research and development, based on the type of market analysis required to identify these, was beyond the scope of the existing resources of the company either in people's or in funds.

The changed texture of the environment was not recognized by an able but traditional management until it was too late. They failed entirely to appreciate that a number of outside events were becoming connected with each other in a way that was leading up to irreversible general change. Their first reaction was to make an herculean effort to defend the traditional product, then the board split on whether or not to make entry into the cheaper unbranded market in a supplier role. Group H.Q. now felt they had no option but to step in, and many upheavals and changes in management took place until a 'redefinition of mission' was agreed, and slowly and painfully the company reemerged with a very much altered product mix and something of a new identity.

FOUR TYPES OF CAUSAL TEXTURE

It was this experience, and a number of others not dissimilar, by no means all of them industrial (and including studies of change problems in hospitals, in prisons, and in educational and political organizations), that gradually led us to feel a need for re-directing conceptual attention to the causal texture of the environment, considered as a quasi-independent domain. We have now isolated four 'ideal types' of causal texture, approximations to which may be thought of as existing simultaneously in the 'real world' of most organizations—though, of course, their weighting will vary enormously from case to case.

The first three of these types have already, and indeed repeatedly, been described—in a large variety of terms and with the emphasis on an equally bewildering variety of special aspects—in the literature of a number of disciplines, ranging from biology to economics and including military theory as well as psychology and sociology. The fourth type, however, is new, at least to us, and is the one that for some time we have been endeavouring to identify. About the first three, therefore, we can be brief, but the fourth is scarcely understandable without reference to them. Together, the four types may be said to form a series in which the degree of causal texturing is increased, in a new and significant way, as each step is taken. We leave as an open question the need for further steps.

Step One

The simplest type of environmental texture is that in which goals and noxiants ('goods' and 'bads') are relatively unchanging in themselves and randomly distributed. This may be called the *placid, randomized environment*. It corresponds to Simon's idea of a surface over which an organism can locomote: most of this is bare, but at isolated, widely scattered points there are little heaps of food (1957, p. 137). It also corresponds to Ashby's limiting case of no connection between the environmental parts (1960, S15/4); and to Schutzenberger's random field (1954, p. 100). The economist's classical market also corresponds to this type.

A critical property of organizational response under random conditions has been stated by Schutzenberger: that there is no distinction between tactics and strategy, 'the optimal strategy is just the simple tactic of attempting to do one's best on a purely local basis' (1954, p. 101). The best tactic, moreover, can be learnt only by trial and error and only for a particular class of local environmental variances (Ashby, 1960, p. 197). While organizations under these conditions can exist adaptively as single and indeed quite small units, this becomes progressively more difficult under the other types.

Step Two

More complicated, but still a placid environment, is that which can be characterized in terms of clustering: goals and noxiants are not randomly distributed but hang together in certain ways. This may be called the *placid, clustered environment*, and is the case with which Tolman and Brunswik were concerned; it corresponds to Ashby's 'serial system' and to the economist's 'imperfect competition.' The clustering enables some parts to take on roles as signs of other parts or become means-objects with respect to approaching or avoiding. Survival, however, becomes precarious if an organization attempts to deal tactically with each environmental variance as it occurs.

The new feature of organizational response to this kind of environment is the emergence of strategy as distinct from tactics. Survival becomes critically linked with what an organization knows of its environment. To pursue a goal under its nose may lead it into parts of the field fraught with danger, while avoidance of an immediately difficult issue may lead it away from potentially rewarding areas. In the clustered environment the relevant objective is that of 'optimal location,' some positions being discernible as potentially richer than others.

To reach these requires concentration of resources, subordination to the main plan, and the development of a 'distinctive competence,' to use Selznick's (1957) term, in reaching the strategic objective. Organizations under these conditions, therefore, tend to grow in size and also to become hierarchical, with a tendency towards centralized control and coordination.

Step Three

The next level of causal texturing we have called the *disturbed-reactive environment*. It may be compared with Ashby's ultra-stable system or the economist's oligopolic market. It is a type 2 environment in which there is more than one organization of the same kind; indeed, the existence of a number of similar organizations now becomes the dominant characteristic of the environmental field. Each organization does not simply have to take account of the others when they meet at random, but has also to consider that what it knows can also be known by the others. The part of the environment to which it wishes to move itself in the long run is also the part to which the others seek to move. Knowing this, each will wish to improve its own chances by hindering the others, and each will know that the others must not only wish to do likewise, but also know that each knows this. The presence of similar others creates an imbrication, to use a term of Chein's (1943), of some of the causal strands in the environment.

If strategy is a matter of selecting the 'strategic objective'—where one wishes to be at a future time—and tactics a matter of selecting an immediate action from one's available repertoire, then there appears in type 3 environments to be an intermediate level of organizational response—that of the *operation*—to use the term adopted by German and Soviet military theorists, who formally distinguish tactics, operations, and strategy. One has now not only to make sequential choices, but to choose actions that will draw off the other organizations. The new element is that of deciding which of someone else's possible tactics one wishes to take place, while ensuring that others of them do not. An operation consists of a campaign involving a planned series of tactical initiatives, calculated reactions by others, and counter-actions. The flexibility required encourages a certain decentralization and also puts a premium on quality and speed of decision at various peripheral points (Heyworth, 1955).

It now becomes necessary to define the organizational objective in terms not so much of location as of capacity or power to move more or less at will, i.e., to be able to make and meet competitive challenge. This gives particular relevance to strategies of absorption and parasitism. It can also give rise to situations in which stability can be obtained only by a certain coming-to-terms between competitors, whether enterprises, interest groups, or governments. One has to know when not to fight to the death.

Step Four

Yet more complex are the environments we have called *turbulent fields*. In these, dynamic processes, which create significant variances for the component organizations, arise from the field itself. Like type 3 and unlike the static types 1 and 2, they are dynamic. Unlike type 3, the dynamic properties arise not simply from the interaction of the component organizations, but also from the field itself. The 'ground' is in motion.

Three trends contribute to the emergence of these dynamic field forces:

1. The growth to meet type 3 conditions of organizations and linked sets of organizations, so large that their actions are both persistent and strong enough to induce autochthonous processes in the environment. An analogous effect would be that of a company of soldiers marching in step over a bridge.
2. The deepening interdependence between the economic and the other facets of the society. This means that economic organizations are increasingly enmeshed in legislation and public regulation.
3. The increasing reliance on research and development to achieve the capacity to meet competitive challenge. This leads to a situation in which a change gradient is continuously present in the environmental field.

For organizations, these trends mean a gross increase in their area of *relevant uncertainty*. The consequences which flow from their actions lead off in ways that become increasingly unpredictable: they do not necessarily fall off with distance, but may at any point be amplified beyond all expectation; similarly, lines of action that are strongly pursued may find themselves attenuated by emergent field forces.

THE SALIENCE OF TYPE 4 CHARACTERISTICS (CASE II)

Some of these effects are apparent in what happened to the canning company of case I, whose situation represents a transition from an environment largely composed of type 2 and type 3 characteristics to one where those of type 4 began to gain in salience. The case now to be presented illustrates the combined operation of the three trends described above in an altogether larger environmental field involving a total industry and its relations with the wider society.

The organization concerned is the National Farmers Union of Great Britain to which more than 200,000 of the 250,000 farmers of England and

Wales belong. The presenting problem brought to us for investigation was that of communications. Headquarters felt, and was deemed to be, out of touch with county branches, and these with local branches. The farmer had looked to the N.F.U. very largely to protect him against market fluctuations by negotiating a comprehensive deal with the government at annual reviews concerned with the level of price support. These reviews had enabled home agriculture to maintain a steady state during two decades when the threat, or existence, of war in relation to the type of military technology then in being had made it imperative to maintain a high level of homegrown food without increasing prices to the consumer. This policy, however, was becoming obsolete as the conditions of thermonuclear stalemate established themselves. A level of support could no longer be counted upon which would keep in existence small and inefficient farmers—often on marginal land and dependent on family labour—compared with efficient medium-size farms, to say nothing of large and highly mechanized undertakings.

Yet it was the former situation which had produced N.F.U. cohesion. As this situation receded, not only were farmers becoming exposed to more competition from each other, as well as from Commonwealth and European farmers, but the effects were being felt of very great changes which had been taking place on both the supply and marketing sides of the industry. On the supply side, a small number of giant firms now supplied almost all the requirements in fertilizer, machinery, seeds, veterinary products, etc. As efficient farming depended upon ever greater utilization of these resources, their controllers exerted correspondingly greater power over the farmers. Even more dramatic were the changes in the marketing of farm produce. Highly organized food processing and distributing industries had grown up dominated again by a few large firms, on contracts from which (fashioned to suit their rather than his interests) the farmer was becoming increasingly dependent. From both sides deep inroads were being made on his autonomy.

It became clear that the source of the felt difficulty about communications lay in radical environmental changes which were confronting the organization with problems it was ill-adapted to meet. Communications about these changes were being interpreted or acted upon as if they referred to the 'traditional' situation. Only through a parallel analysis of the environment and the N.F.U. was progress made towards developing understanding on the basis of which attempts to devise adaptive organizational policies and forms could be made. Not least among the problems was that of creating a bureaucratic elite that could cope with the highly technical long-range planning now required and yet remain loyal to the democratic values of the N.F.U. Equally difficult was that of developing mediating institutions—agencies that would effectively mediate the relations between agriculture and other economic sectors without triggering off massive competitive processes.

These environmental changes and the organizational crisis they induced were fully apparent two or three years before the question of Britain's possible entry into the Common Market first appeared on the political agenda—which, of course, further complicated every issue.

A workable solution needed to preserve reasonable autonomy for the farmers as an occupational group, while meeting the interests of other sections of the community. Any such possibility depended on securing the consent of the large majority of farmers to placing under some degree of N.F.U. control matters that hitherto had remained within their own power of decision. These included what they produced, how and to what standard, and how most of it should be marketed. Such thoughts were anathema, for however dependent the farmer had grown on the N.F.U. he also remained intensely individualistic. He was being asked, he now felt, to redefine his identity, reverse his basic values, and refashion his organization—all at the same time. It is scarcely surprising that progress has been, and remains, both fitful and slow, and ridden with conflict.

VALUES AND RELEVANT UNCERTAINTY

What becomes precarious under type 4 conditions is how organizational stability can be achieved. In these environments individual organizations, however large, cannot expect to adapt successfully simply through their own direct actions—as is evident in the case of the N.F.U. Nevertheless, there are some indications of a solution that may have the same general significance for these environments as have strategy and operations for types 2 and 3. This is the emergence of *values that have overriding significance for all members of the field*. Social values are here regarded as coping mechanisms that make it possible to deal with persisting areas of relevant uncertainty. Unable to trace out the consequences of their actions as these are amplified and resonated through their extended social fields, men in all societies have sought rules, sometimes categorical, such as the ten commandments, to provide them with a guide and ready calculus. Values are not strategies or tactics; or Lewin (1936) has pointed out, they have the conceptual character of 'power fields' and act as injunctions.

So far as effective values emerge, the character of richly joined, turbulent fields changes in a most striking fashion. The relevance of large classes of events no longer has to be sought in an intricate mesh of diverging causal strands, but is given directly in the ethical code. By this transformation a field is created which is no longer richly joined and turbulent but simplified and relatively static. Such a transformation will be regressive, or constructively adaptive, according to how far the emergent values adequately represent the new environmental requirements.

Ashby, as a biologist, has stated his view, on the one hand, that examples of environments that are both large and richly connected are not common, for our terrestrial environment is widely characterized by being highly subdivided (1960, p. 205); and, on the other, that, so far as they are encountered, they may well be beyond the limits of human adaptation, the brain being an ultrastable system. By contrast the role here attributed to social values suggests that this sort of environment may in fact be not only one to which adaptation is possible, however difficult, but one that has been increasingly characteristic of the human condition since the beginning of settled communities. Also, let us not forget that values can be rational as well as irrational and that the rationality of their rationale is likely to become more powerful as the scientific ethos takes greater hold in a society.

MATRIX ORGANIZATION AND INSTITUTIONAL SUCCESS

Nevertheless, turbulent fields demand some overall form of organization that is essentially different from the hierarchically structured forms to which we are accustomed. Whereas type 3 environments require one or other form of accommodation between like, but competitive, organizations whose fates are to a degree negatively correlated, turbulent environments require some relationship between dissimilar organizations whose fates are, basically, positively correlated. This means relationships that will maximize cooperation and which recognize that no one organization can take over the role of 'the other' and become paramount. We are inclined to speak of this type of relationship as an *organizational matrix*. Such a matrix acts in the first place by delimiting on value criteria the character of what may be included in the field specified—and therefore who. This selectivity then enables some definable shape to be worked out without recourse to much in the way of formal hierarchy among members. Professional associations provide one model of which there has been long experience.

We do not suggest that in other fields than the professional the requisite sanctioning can be provided only by state-controlled bodies. Indeed, the reverse is far more likely. Nor do we suggest that organizational matrices will function so as to eliminate the need for other measures to achieve stability. As with values, matrix organizations, even if successful, will only help to transform turbulent environments into the kinds of environment we have discussed as 'clustered' and 'disturbed-reactive'. Though, with these transformations, an organization could hope to achieve a degree of stability through its strategies, operation, and tactics, the transformations would not provide environments identical with the originals. The strategic objective in the transformed cases could no longer be stated simply in terms of optimal location (as in type 2) or capabilities (as in type 3). It must now rather be formulated in terms of *institutionalization*. According to Selznick (1957) organizations become institutions through the embodiment of organizational values which relate them to the wider society.[1] As Selznick has stated in his analysis of leadership in the modern American corporation, 'the default of leadership shows itself in an acute form when *organizational* achievement or survival is confounded with *institutional* success' (1957, p. 27). '. . . the executive becomes a statesman as he makes the transition from administrative management to institutional leadership' (1957, p. 154).

The processes of strategic planning now also become modified. In so far as institutionalization becomes a prerequisite for stability, the determination of policy will necessitate not only a bias towards goals that are congruent with the organization's own character, but also a selection of goal-paths that offer maximum convergence as regards the interests of other parties. This became a central issue for the N.F.U. and is becoming one now for an organization such as the National Economic Development Council, which has the task of creating a matrix in which the British economy can function at something better than the stop-go level.

Such organizations arise from the need to meet problems emanating from type 4 environments. Unless this is recognized, they will only too easily be construed in type 3 terms, and attempts will be made to secure for them a degree of monolithic power that will be resisted overtly in democratic societies and covertly in others. In the one case they may be prevented from ever undertaking their missions; in the other one may wonder how long they can succeed in maintaining them.

An organizational matrix implies what McGregor (1960) has called Theory Y. This in turn implies a new set of values. But values are psycho-social commodities that come into existence only rather slowly. Very little systematic work has yet been done on the establishment of

---

[1]Since the present paper was presented, this line of thought has been further developed by Churchman and Emery (1964) in their discussion of the relation of the statistical aggregate of individuals to structured role sets: "Like other values, organizational values emerge to cope with relevant uncertainties and gain their authority from their reference to the requirements of larger systems within which people's interests are largely concordant."

new systems of values, or on the type of criteria that might be adduced to allow their effectiveness to be empirically tested. A pioneer attempt is that of Churchman and Ackoff (1950). Likert (1961) has suggested that, in the large corporation or government establishment, it may well take some ten to fifteen years before the new type of group values with which he is concerned could permeate the total organization. For a new set to permeate a whole modern society the time required must be much longer—at least a generation, according to the common saying—and this, indeed, must be a minimum. One may ask if this is fast enough, given the rate at which type 4 environments are becoming salient. A compelling task for social scientists is to direct more research onto these problems.

SUMMARY

(a) A main problem in the study of organizational change is that the environmental contexts in which organizations exist are themselves changing—at an increasing rate, under the impact of technological change. This means that they demand consideration for their own sake. Towards this end a redefinition is offered, at a social level of analysis, of the causal texture of the environment, a concept introduced in 1935 by Tolman and Brunswik.

(b) This requires an extension of systems theory. The first steps in systems theory were taken in connection with the analysis of internal processes in organisms, or organizations, which involved relating parts to the whole. Most of these problems could be dealt with through closed-system models. The next steps were taken when wholes had to be related to their environments. This led to open-system models, such as that introduced by Bertalanffy, involving a general transport equation. Though this enables exchange processes between the organism, or organization, and elements in its environment to be dealt with, it does not deal with those processes in the environment itself which are the determining conditions of the exchanges. To analyse these an additional concept—the causal texture of the environment—is needed.

(c) The laws connecting parts of the environment to each other are often incommensurate with those connecting parts of the organization to each other, or even those which govern exchanges. Case history I illustrates this and shows the dangers and difficulties that arise when there is a rapid and gross increase in the area of relevant uncertainty, a characteristic feature of many contemporary environments.

(d) Organizational environments differ in their causal texture, both as regards degree of uncertainty and in many other important respects.

A typology is suggested which identifies four 'ideal types', approximations to which exist simultaneously in the 'real world' of most organizations, though the weighting varies enormously:

1. In the simplest type, goals and noxiants are relatively unchanging in themselves and randomly distributed. This may be called the placid, randomized environment. A critical property from the organization's viewpoint is that there is no difference between tactics and strategy, and organizations can exist adaptively as single, and indeed quite small units.
2. The next type is also static, but goals and noxiants are not randomly distributed; they hang together in certain ways. This may be called the placid, clustered environment. Now the need arises for strategy as distinct from tactics. Under these conditions organizations grow in size, becoming multiple and tending towards centralized control and coordination.
3. The third type is dynamic rather than static. We call it the disturbed-reactive environment. It consists of a clustered environment in which there is more than one system of the same kind, i.e., the objects of one organization are the same as, or relevant to, others like it. Such competitors seek to improve their own chances by hindering each other, each knowing the others are playing the same game. Between strategy and tactics there emerges an intermediate type of organizational response—what military theorists refer to as operations. Control becomes more decentralized to allow these to be conducted. On the other hand, stability may require a certain coming-to-terms between competitors.
4. The fourth type is dynamic in a second respect, the dynamic properties arising not simply from the interaction of identifiable components systems but from the field itself (the 'ground'). We call these environments turbulent fields. The turbulence results from the complexity and multiple character of the causal interconnections. Individual organizations, however large, cannot adapt successfully simply through their direct interactions. An examination is made of the enhanced importance of values, regarded as a basic response to persisting areas of relevant uncertainty, as providing a control mechanism, when commonly held by all members in a field. This raises the question of organizational forms based on the characteristics of a matrix.

(e) Case history II is presented to illustrate problems of the transition from type 3 to type 4. The perspective of the four environmental types is used to clarify the role of Theory X and Theory Y as representing a trend

in value change. The establishment of a new set of values is a slow social process requiring something like a generation—unless new means can be developed.

REFERENCES

ASHBY, W. ROSS (1960). *Design for a brain*. London: Chapman & Hall.

BARKER, R. G. & WRIGHT, H. F. (1949). Psychological ecology and the problem of psychosocial development. *Child Development* **20**, 131-43.

BERTALANFFY, L. von (1950). The theory of open systems in physics and biology. *Science* **111**, 23-9.

CHEIN, I. (1943). Personality and typology. *J. Soc. Psychol.* **18**, 89-101.

CHURCHMAN, C. W. & ACKOFF, R. L. (1950). *Methods of inquiry*. St. Louis: Educational Publishers.

CHURCHMAN, C. W. & EMERY, F. E. (1964). On various approaches to the study of organizations. Proceedings of the International Conference on Operational Research and the Social Sciences, Cambridge, England, 14-18 September 1964. Published in book form as *Operational research and the social sciences*. London: Tavistock Publications, 1965.

HEYWORTH, LORD (1955). *The organization of Unilever*. London: Unilever Limited.

LEWIN, K. (1936). *Principles of topological psychology*. New York: McGraw-Hill.

LEWIN, K. (1951). *Field theory in social science*. New York: Harper.

LIKERT, R. (1961). *New patterns of management*. New York, Toronto, London: McGraw-Hill.

McGREGOR , D. (1960). *The human side of enterprise*. New York, Toronto, London: McGraw-Hill.

PEPPER, S. C. (1934). The conceptual framework of Tolman's purposive behaviorism. *Psychol. Rev.* **41**, 108-33.

SCHUTZENBERGER, M. P. (1954). A tentative classification of goal-seeking behaviours. *J. Ment. Sci.* **100**, 97-102.

SELZNICK, P. (1957). *Leadership in administration*. Evanston, Ill.: Row Peterson.

SIMON, H. A. (1957). *Models of man*. New York: Wiley.

TOLMAN, E. C. & BRUNSWIK, E. (1935). The organism and the causal texture of the environment. *Psychol. Rev.* **42**, 43-77.

Open Systems Planning

G. K. Jayaram

Open Systems Planning (OSP) entails the identification and delineation of the (person-task-process-environment) dynamics. In other words, the model highlights the vital need to understand the external and internal environments (of the system), the basic core socio-technical process/processes (of the system), the dynamic equilibrium in which the transactions exist between and amongst the internal and external domains as well as across the boundaries (of the system) and the extension of all these three dimensions into the future through both the (relatively) subjective and objective filters of the planning group assembled to work through the model. This can only be done in a "community of trust," which emphasizes the prerequisites of sharing and trusting processes. If the OSP exercise is not imbued with a considerable degree of trust, the product would be mere superficialities or, worse, distortions in a game of political oneupmanship. Even without the games of power, it would still be a shell without spirit or substance, a mere symbol without any significance.

These features evidence that this model harvests the benefit of learning from all the earlier phases and attempts to combine the best of the heritage. It is, in other words, not an idiosyncratic special tool, but a first faltering step at the integration of the learnings from the major contributions of the past.

Reprinted from G. K. Jayaram, "Open Systems Planning." In W. G. Bennis, K. D. Benne, R. Chin, & K. Corey (Eds.), *The Planning of Change* (3rd ed.), pp. 275-283. New York: Holt, Rinehart and Winston, 1976. Used by permission of the author.

The process of open systems planning* consists of the following steps:

1. Creation of the "Present Scenario":
   a. External domains of the environment—expectations and interactions.
   b. Internal identity, expectations and interactions;
   c. Transactions across the boundaries of the system.
2. Creation of "Realistic Future Scenario" (under a, b & c of (1) above).
3. Creation of "Idealistic Future Scenario" (IFS) under a, b, & c of (1) above.
4. Sharing of IFS's, comparison of (1), (2) & (3) and identification of broad areas of consensus, controversy and total disagreement about future.
5. Temporal (time-based) planning of action programs for the areas identified earlier.

The above five steps are described in what follows. Interspersed through the description are examples or notes from personal consulting experience relevant at that particular point.

DESCRIPTION

1. Creation of "Present Scenario"

(a) External The group as a whole is asked to concentrate on a concrete socio-technical system central to all the members; e.g., the factory or the ship or the school or the hospital, and enumerate all the *expectations* coming on that system from the external environment. [Figure 1].

Note: The rationale behind the choice of 'System' for focus should be that it should reflect common stakes for all the subsystems of the group.

Note: The exercise starts with looking at the external environment and *not* with an analysis of the internal milieu. This is a deliberate choice dictated both by concepts and experience. Conceptually, the crucial insights involved in thinking of one's system as part of many

---

*The description of the workshop model which follows is adopted from the manuscript written by the author under the title "Major Field Statement" in 1970 and another manuscript in 1972 at UCLA. Changes and modifications from the earlier versions appear here, as a result of four years of usage of the model in consulting practice in diverse types of organizations. The coauthor of the model is Mr. Charles Krone; however, the responsibility for the commissions or omissions in this and earlier descriptions of the model rest entirely with the author.

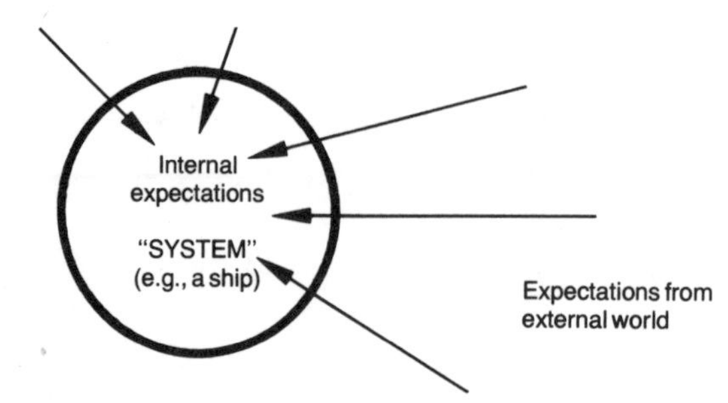

Figure 1.

larger systems with simultaneous multiple membership and engaged in a dynamic process of give and take with the environmental domains are most valuable starting points for holistic analysis. Experience with the model indicates that the 'view of the navel'—the delineation of the internal being difficult and seductive at the same time—permits a group to wallow in the egocentric, ptolemaic pursuit of one's own tail. It is tougher to turn from inside to outside than the other way around. Besides, the habit pattern, which is the exclusive attention to the internal system, needs to be broken.

Note: Part of the 'technology' of this process consists of writing in pictures and words the stream of what is being expressed by the group on newsprint sheets in different colored crayons and pasting all these data on the walls of the room. The consultant may act as the synthesizer and recorder of data at times and may encourage the group to do so at other times. As high a degree of creativity as the group can muster should be evoked in this whole process, including use of recording materials, writing, use of idioms—metaphors (the system's own mythology, language, symbols and rituals).

This is an unstructured, open-ended process. The consultant acts as recorder. He encourages participation from all around the room at random. He deliberately encourages metaphorical phrases for the expectations so that the scenario may come alive for the members.

When this process is done sufficiently, according to the discretion of the group and the consultant, the group is asked to classify the expectations according to the sources from which they are *perceived to* emanate—i.e., domains of expectation (reference groups/role senders, etc.).

In other words, the first question is, "What is expected of this system?" The second question is, "Who expects this?"

This process of classification by domains may make the client group realize that there are many relevant domains left out from the earlier data and/or that there are many expectations felt for which no source is traceable and/or there are some obvious domains whose expectations are not clear.

At this stage, they may also begin to perceive shared expectations among many domains, unshared ones, contradictory ones. They may begin to glimpse the dynamic interaction among the variables in between the domains.

All these perceptions may persuade them to redo the first scenario with greater sophistication and insight a second time around.

Note: This step by itself may have the following constructive results:

(a) This may be the first time the different functionaries are sharing in a real (and not gamey) sense, their individual perceptions about the world 'out there'. This can be a revealing 'ah ha' experience if the group members are open to change from one another.

Example: One of the early groups who experienced this process had their first major breakthrough at this stage. Two different departmental heads sharing their perceptions of the external expectations, realized together that each had come to view the other as at best dumb, at worst an enemy, because each saw the other doing things contrary to one's own perceptions of the pressure from outside. Each one realized here for the first time that it was a different (and contradictory) perception of the expectations from the top management outside that led each, in a loyalty to the system, to do what he or she did. The sense of relief and unity in the room was electric.

(b) Internal environment—its identity and core processes The objective of this step is for the group to identify and represent the real picture of their system from its center to its peripheries in all its complexity and richness. The step can be partially similar in the questions asked to the step gone through earlier in the case of the external environment—i.e., the internal expectations, domains, interactions and values.

Members of the group are encouraged to range, in their search from intra- and inter-personal to inter-subsystemic expectations. (Needless to say, the discretion of the consultant is to be exercised in narrowing or broadening the band of search, depending on the political climate in the group, the stage of cohesiveness, interpersonal-intergroup trust, structural and functional needs of the system, etc.)

Note: This step deserves a special note since client groups seem to experience great difficulty in responding to the questions at this stage. The real question which is being asked here is a search for identity—systematic or organizational identity. This can be conceived both as content and process, i.e., identity, and core process or processes which express that identity. It is this crucial and intimate nature of this question that seems to cause the blocks in the creativity of the group response. In fact, in contrast to this, the earlier step of delineating the external environment is done relatively easily since it is more distant, seemingly less threatening and amenable to greater objectivity. The pains surrounding the attempt to create the internal scenario reminds one of the ancient Indian saying: "The shadow is the darkest underneath the candle"—self-perception and awareness is difficult at the organizational and institutional level as much as at the individual personality level.

To facilitate the creative process, the consultant may suggest such questions as: "What would you give up if you had to give up 90% of what you are, as a system?" He may also use such conceptual categories as activities, interactions and sentiments to help the group analyse the internal milieu. He may ask the same set of questions as were asked about the external environment. He may also encourage them to identify where the 'tension tendons' or energy of the system lie. The purpose of the intervention should not be to escape the rigour of the question, but to approach it from various sides like approaching a fortress, looking for a natural entry into the 'systemic imperatives'. This may be different for different systems.

Example: The plant management of a large plant in Southern California struggled for an interminable amount of time with this question. They wondered aloud whether they should describe the organizational chart or departmental set-up or functions or profit centers or cost centers. The traditional pictures paraded as 'the organization' in front of outsiders were considered and discarded as lacking 'soul'. None seemed to capture where the energy of the system lay. At long last, they began a search of what internal domains seemed to matter. What they came up with were a set of dichotomies or 'tension bipolarities' such as young vs. old, women vs. men, Black vs. white, mechanics vs. non-mechanics, crafts vs. unskilled, etc. This was their breakthrough towards opening the doors of data on what the internal life of the system was like for the insiders.

This stage may reveal the host of borderline expectations, which cannot be easily labeled "internal" or "external." The individual himself is always a part of his own environment and is partly responsible for it (Menninger). This is equally true for the system. The group may go through similar insights, as in the case of external domains, in the case of internal subsystems. In addition, they may begin to glimpse transactions across the boundaries between the internal parts and the external domains, the degree of permeability across different parts of the boundary and the complex networks of relationships among variables in all the dimensions.

(c) The consultant now responds to the feeling of overwhelming complexity by introducing the notion of value systems underlying the expectations. (Incidentally, none of these terms, "value systems" included, need be elaborated upon in their technical context; instead, their colloquial meanings would suffice to produce effective process.)

The group is asked the question, "Why does who expect what?" The value systems perceived to be prevalent in each domain leading to expectations are traced and recorded, again with emphasis on striking anecdotes, phrases, etc. Similar delineation of the nuances of the value systems in the internal world is done, starting from each one's personal values as reflected in the organizational context.

Note: Invariably at this stage of the process, there develops in the group a deep sense of gloom. Approximately a day long search has taken place to weave the present scenario of the system's outside and inside. A pervasive feeling of helplessness and uncontrollability about the system's destiny is manifest at this point. This may get expressed in various overt or covert ways, including projection of sense of failure to the consultant. The group (or some members) may turn with varying degrees of hostility and rejection on the consultant and say: "So what! Big deal. We knew all of this before we started" or "Now that we know how bad it all is, what have you done to help us get through it? Nothing!" If this deep gloom and reaction is reminiscent of similar phenomena in T-groups or therapeutic interventions, it is not a coincidence but the evidence of isomorphic existential realities at different systemic levels and is conceptually warranted. One can only suggest (to the consultant) to stay with it—to hang in there and let the group do the same, eschewing the temptation to fly from the pain of the overwhelming complexity of the scenario and a perception of one's finite capacity to deal with it.

2. Creation of "Realistic Future Scenarios"

The group is now asked: "Suppose there was no deliberate intervention for any kind of change. What would the system scenario be in some indeterminate future?" (If the curve from past to present is projected on to the future [same slope] where would the system be at some point in the future?)

Calling this *"The Realistic Future Scenario,"* the group is asked to go through the same steps as in the creation of the present scenario, ending again with the projection, to their logical consequences, of their present value systems.

Note: The purpose in introducing this step is as follows: The seeds of the future may be with us at the present moment; but it is almost universally true that it is difficult to perceive the future or the futures that are inherent or emerging from the present. Some reasons for this myopia may be obvious and some subtle. It may be part of the human condition to suffer "the tyranny of the eternal yesterday" (Weber). The yesterdays are never so dead that their ghosts won't haunt to prevent the unborn tomorrows to cast their prenatal shadows. Also the capacity to 'separate the wheat from the chaff' in the present—the mere fad from the deeper step-function or quantum change, the peripheral from the central—is surely lacking in most groups (including Social Sciences). Hence the attempt to project the scenario to a point in the future may help as a start for the distinguishing of what may be here to stay from what may be sure to pass. Also it may bring to the surface suppressed feelings (positive or negative) about present trends. This may energize action about those trends.

This step, in experience, has proved to be of varying importance for different groups.

3. Creation of "Idealistic Future Scenarios"

At the third stage, the group is asked the question: "Suppose you made change interventions, what *alternative futures* would you wish *to create*?" or "If you had the power to change anything in the earlier scenarios, *what* would you like to change and *how* would the scenario look when changed?"

For this process the group is asked to split into dyads or triads—with the choice of membership of each group either made from within the group or by the consultant (if he has, by now, special conceptual constructs about the nature of the group and feels optimal to prescribe certain composition in each group).

Note: This is the stage to dream unfettered by realities—temporarily at least. It is to be noted that the question does *not* include "how or why would you make changes?" Self-conscious creativity is encouraged in the small subgroups. The consultant can float around or leave the process uninterrupted. It is to be expected that individuals will bring forth their biases, which may hail from their functional specializations, political orientations, socio-economic backgrounds, cultural-ethnic factors or idiosyncratic personality features. It is valuable to encourage such subjective creation and discussion in the subgroup.

4. Sharing of IFS's

(a) Now, all the small groups are invited to gather back in one large group and present to one another their *idealistic future scenarios*.

In this process, the groups may be reminded to perceive *all the relevant dimensions* of the causal texture of the environment—socio-technical, socio-psychological, politico-legal, economic, ecological, etc., and the impact of the changes in these dimensions on their future designs. Each individual is to ask himself and others the question: "If the suggested changes were to take place, how would they affect anything and/or everything else (in terms of quality and quantity)?" Again, the choice of introducing the appropriate degree of complexity into the process is left to the discretion of the consultant and the group—the reading of the pulse of the group.

(b) All the previous scenarios are placed in front of the group. The question is: "What are the variances you see between (a) the present and the idealistic futures, (b) the realistic and idealistic futures and why. Speculate. "Can you do anything about absorbing the variance with sufficient lead time? If so, what? If not, why not?" The list of questions at this stage, with the wealth of data and the possible absorption of the client-consultant system into the process, is only limited by the limits of collective comprehension and imagination.

(c) As a natural product of the two steps described above in (a) and (b), the group is asked to identify *three sets of issues* for the future.

(i) areas of broad concurrence or consensus among the group (as evidenced in the above steps including the Future Scenarios), called "the Yes list" (i.e., "we are all agreed");

(ii) areas of uncertainty, partial agreement or controversy (the reasons for which may be lack of data in the group or lack of conviction or willingness or values to face those issues), called "the maybe list" (i.e., "we are doubtful");

(iii) areas of intense disagreement (the reasons ranging all over the broad—philosophical, contextual or emotional—bases), called "The Yes-No" list.

5. Temporal Planning

The group is asked at this final stage facing the three lists: "What would you do about each item on each list in three time dimensions:

(i) tomorrow;

(ii) six months from now;

(iii) two years from now."

The group has to decide on the choice of change interventions they would/could/should make. Here again, the heart of the process is the delineation of values underlying each choice of intervention—why to intervene at all? What kind of intervention and why? Where to intervene and why there? Why proceed towards that particular future?

The nature of action steps would differ in the case of the three lists. The "yes-list" may invite concrete planning and implementation process of the details of each issue, since issues have been identified and consensus reached. In the case of the "Yes-No" list, processes of unearthing the causes of the sharp disagreements may be at any of the following five levels: ideology, values, strategy, operation, tactics.[1] The "maybe" list may need further research on issues or participation of relevant people in the data search or clarification. The action steps here may include such processes as the replication of open systems planning at other levels of the system and/or information gathering processes. As the final and important step, the group decides on the follow-up schedule to (a) share the flow of action steps and (b) update the scenarios.

FEATURES TO KNOW ABOUT THE PROCESS

1. OSP is effective as an iterative process. Since the data (both facts and values) are distributed all over the open system, the exercise (in its data-gathering function) can be used laterally and vertically all over the organization.

2. The primary or initial group to walk through the model preferably should be the formal planning group of the organization. Without the feeling that the group has the power and legitimacy to perform the task, the power of the model in its execution is decreased. However, this group may decide the need for particular other groups or individuals to participate in a similar process.

3. There is vital need to follow up on this process at regular frequency. It becomes in a sense like the board on which the movement of stocks is indicated in the Stock Exchange. Only here the changes to be posted are a lot more complex, a lot less certain, but equally (or more) unpredictable—hence the great need to prepare the framework and update it regularly. Some issues may recur at different points of time at different intensities or new issues may enter the picture, changing the scenarios substantially. The issues may arise in the external or the internal or some combination of both.

4. There needs to exist, as a precondition for an effective open systems planning process, a degree of trust which permits open communication of positive and negative interpersonal aspects. This has been mentioned before. The design implication of this feature is

[1]From G. K. Jayaram, "Conflict Resolution: Some Concepts" and "Fact-Value Analysis: A Method for Conflict Resolution." (Unpublished monographs.)

thusly: The intervener may diagnose mutually with the client group (by whatever diagnostic means available or preferable in the context, e.g., depth interviews, survey questionnaires, observation of group meetings, etc.) the level of mutual trust in the group. He may then decide on a 'team-building exercise' as a prerequisite before OSP *or* he may decide that an adequately open expression exists in the group for the sharing of cognitive-affect data needed in OSP.

Figure 2 indicates that an alternating between the interpersonal openness processes team-building and systemic openness processes like OSP may be advisable. At no level (individual, interpersonal, group or system) should it be assumed that a one-shot process suffices 'to hold the glue' forever. Like hunger and thirst and sex at the individual level, the systemic needs at the process and content levels require responses at regular frequencies.

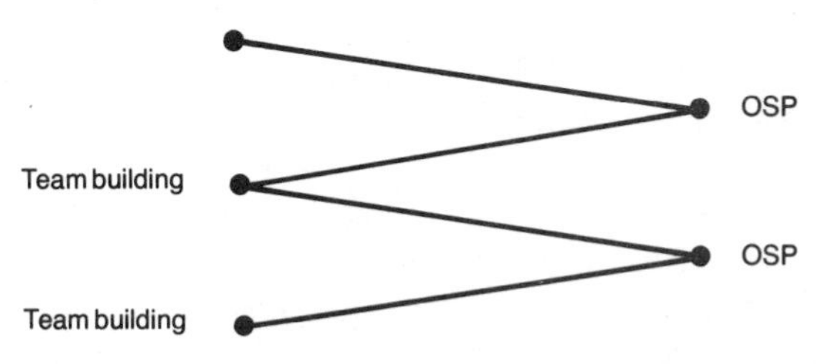

Figure 2.

There need be no anxiety to get all the data in one shot, since this is an iterative design, involving successively larger or smaller chunks of organizational space in the spatio-temporal dimensions.

5. The consulting resources would do well to possess considerable skills (diagnostic and training skills) at the process, structure and cognitive (conceptual) content levels.

6. The model is useful in any kind of system ranging from family through organizations (industrial and non-industrial), communities, etc.

7. Finally, a note on the time duration of the workshop:

Note: The group would work for an initial one- and one-half days. The steps that would be completed in this period are 'the present scenario', 'the realistic future scenario', and 'the idealistic future scenario'. There would be a specific contract suggested right at the start to the effect that as they work through these steps, the group would deliberately list 'areas of incomplete knowledge' in any domain, internal or external. At the end of 1½ days, the group takes the responsibility to research, by whatever means necessary and feasible, these items of incomplete knowledge. The time allotted would be one week. This would also be the time for individual reflection on what has transpired in the initial sessions. At the end of the week, the group gathers for the second and final one- and one-half days of work. They will start out by sharing both their research and their reflections with one another. The products of the earlier sessions would be modified, if needed, according to new data. This will be for 2-3 hours. Then the group would start on the last phase of the model—'the temporal planning.'

Hence the actual time needed for OSP is three days, but this duration is cut in two allowing for a very valuable week of research and reflection.

FINAL COMMENTS

This design is a technique for appreciation and for instrumentation. It helps the client-system evolve, from within its own self (and in collaboration with a consultant at the initial stages), the Scenarios of its present world and its probable future worlds. Both the internal and external textures in each of these temporal dimensions can be delineated, with great richness of detail and as organized complexities. The client creates his own crystal ball, develops eyes that can gaze into the ball and looks into change, if necessary, to know, at all cost, in order to be prepared for the holocaust—or the utopia.

At successive stations in the process, the client comes to evoke the underlying value systems in his internal environment and attempts to empathize with the value systems of the domains relevant to him in the external environment. This process will be seen by the client to involve considerable fact-search and feeling-search, not only among the particular groups involved in this process, but also in other relevant groups.

This data search and a gradual filling of the canvas—the creation of the three scenarios—at its minimum, is a data-generating technique and at its best an active intervention for change.

PART III:

SOCIOTECHNICAL SYSTEM THEORY

Introduction

The articles in this section are more concerned with the organization's internal social and technological systems than with its environment. As Trist says in his article, "On Sociotechnical Systems," open system and sociotechnical thinking imply one another. Obviously, although we can concentrate on a firm's internal dynamics, we always must consider the impact of the environment on the conclusions drawn.

In contrast to the preceding articles, the following articles cover more than open system social analysis. Additionally, the authors develop a concept of the firm as a dynamic interaction of social and technological forces. Trist reports learnings from actual sociotechnical system interventions. He makes the point that although the technology employed by an organization sets limits on acceptable social behaviors, there nevertheless is some degree of choice in the structure of social systems. The objective, therefore, is to find the social and technological arrangement that meets both the needs of employees in terms of the quality of their working life and organizational goals. Trist also notes that any such "jointly optimized system" will have a short tenure if the organization does not place itself in its environment in such a way as to assure its continued well-being.

Bucklow describes semi-autonomous work units that have evolved from sociotechnical system interventions. She notes that responsibly autonomous groups have yielded greater performance and greater worker satisfaction and that the essential work performed in organizations, the "primary task," is the key concept that integrates the technological, economic, and sociopsychological aspects of the system. For this reason, a real transference of power and control to the group for the primary task has successfully motivated employees and efficiency has increased.

Cherns gives an up-to-date account of "The Principles of Sociotechnical Design." These include: making the sociotechnical design compatible with the organization's goals; providing minimal specification of tasks; controlling production variances as near to their source as possible; allowing necessary development by providing multifunctional capacity; basing the boundaries of the system on time, territory, or technology; providing information flow designing support systems to reinforce the behaviors that the organizational structure is designed to elicit; providing a high quality of work life; and allowing the design to change when necessary to meet environmental demands. The paper is well written, very readable, and should be of great value to those who have little prior background in sociotechnical system theory.

With some conceptual tools in mind, the reader is prepared to diagnose and effect change in a sociotechnical system. (Guidelines for diagnosis and change will be presented in Part IV.)

On Socio-Technical Systems

E. L. Trist

My aim in this paper is to present a frame of reference within which industrial enterprises may be studied empirically—whether as wholes or as parts. This frame of reference has developed over a number of years between my colleagues and myself at the Tavistock Institute and has entailed a shift from looking at enterprises as closed social systems to looking at them as open socio-technical systems. That is to say, there has been a shifting on the one hand from thinking in terms of closed systems to thinking in terms of open systems; on the other there has been a change from a point of view in which enterprises were considered solely as social systems to one in which the technology is also taken into account and an attempt is made to relate the social and technological systems to each other. These relationships may of course be studied at any level: that of the individual, the primary work group, larger internal units involving various levels of management, and the enterprise as a whole. Several of us—my colleagues F. E. Emery and A. K. Rice, for example, and certainly I myself—have also come to believe that open system and socio-technical thinking imply each other in the study of the enterprise. If in this field of work one starts on a piece of research socio-technically, sooner or later one finds one's self using open system theory, implicitly or explicitly. Similarly, if one's original approach to a problem is in terms of open system theory, our finding at any rate is that one ends up with a socio-technical rather than a purely social analysis.

Reprinted from E. L. Trist, "On Socio-Technical Systems," an open university lecture jointly sponsored by the Departments of Engineering and Psychology at the University of Cambridge, 18th November, 1959. Used by permission.

What I should like to see develop therefore is a general theory of the enterprise as an open socio-technical system—the present paper is offered as a contribution towards this end. The theoretical treatment follows closely that adopted in a joint paper by Emery and myself given to the Sixth International Congress of the Institute of Management Sciences last September, but much more research data will be presented than was possible on that occasion.

The analysis of the characteristics of enterprises as systems has, I would say, strategic significance for furthering our understanding of a great number of specific industrial problems. The more we know about these systems, the more we are able to identify what is relevant to a particular problem and to detect problems that tend to be missed by the conventional framework of problem analysis.

The value of studying enterprises as systems has been demonstrated in a series of empirical studies. Many of these studies have been informed by a broadly conceived concept of bureaucracy, derived from Weber and influenced by Parsons and Merton. These studies have of course conceived industrial organizations as social rather than socio-technical systems. The early Tavistock work is no exception to this.

Granted the importance of system analysis, and before considering social v. socio-technical, there arises the prior question of whether an enterprise should be construed as a 'closed' or an 'open system', i.e. relatively 'closed' or 'open' with respect to its external environment. Von Bertalanffy first introduced this general distinction in contrasting biological and physical phenomena, one of the best accounts of his work from the standpoint of the social scientist being his 1950 paper on "The Theory of Open Systems in Physics and Biology", published in the journal *Science*. In the realm of social theory, however, there has been a strong tendency to continue thinking in terms of a 'closed' system, that is to regard the enterprise as sufficiently independent to allow most of its problems to be analysed with reference to its internal structure and without reference to its external environment. As a first step, closed system thinking has been fruitful, in psychology and industrial sociology, in directing attention to the existence of structural similarities, relational determination and subordination of part to whole. However, it has tended to be misleading on problems of growth and the conditions for maintaining a 'steady state'. The formal physical models of 'closed systems' postulate that, as in the second law of thermodynamics, the inherent tendency is to grow toward maximum homogeneity of the parts and that a steady state can only be achieved by the cessation of all activity. In practice system theorists in social science (and these include such key anthropologists as Radcliffe-Brown) have refused to recognize these implications. They have instead, but by the same token, tended to focus on the statics of social structure and to neglect the study of structural change. In an attempt to overcome this bias, Merton has suggested that "the concept of dysfunction—which implies strain, stress and tension on the structural level—provides an analytical approach to the study of dynamics and change". This concept has been widely accepted. But, while it draws attention to sources of imbalance within an organization, it does not conceptually reflect the mutual permeation of an organization and its environment that is the cause of such imbalance. It still retains the limiting perspective of 'closed system' theorizing.

The alternative conception of 'open systems' carries the implication that such systems may spontaneously re-organize towards states of greater heterogeneity and complexity, and that they achieve a 'steady state' at a level where they can still do work. Enterprises would appear to possess such 'open system' characteristics. They grow by processes of internal elaboration. They manage to achieve a steady state while doing work. They achieve a quasi-stationary, equilibrium in which the enterprise as a whole remains constant, with a continuous 'throughout', despite a considerable range of external changes.

The appropriateness of the concept of 'open system' can be settled, however, only by examining in some detail what is involved in an enterprise achieving a steady state. The continued existence of any enterprise presupposes some regular exchange in products or services with other enterprises, institutions and persons in its external social environment. If it is going to be useful to speak of steady states in an enterprise, they must be states in which this exchange is going on.

Now the conditions for regularizing this exchange lie both within and without the enterprise. Internally, this presupposes that an enterprise has at its immediate disposal the necessary material supports for its activities—a workplace, materials, tools and machines—and, no less, a work force able and willing to make the necessary modifications in the material 'throughout' or provide the requisite supports and to organize the actions of its human agents in a rational and predictable manner. Externally, the regularity of exchange or commerce with the environment may be influenced by a broad range of independent external changes affecting alike markets for products and inputs of labour, materials and technology.

If we examine the factors influencing the ability of an enterprise to maintain a steady state in the face of these broader environmental influences we find *first, with regard to outputs or exports* that the variation in the output markets that can be tolerated without structural change is a function of the flexibility of the technical productive apparatus—its ability to vary its rate, its end product or the mixture of its products. Variation in the output markets may itself be considerably reduced by a display of distinctive competence. Thus the output markets will be more attached to a given enterprise if it has, relative to other producers, a distinctive competence—a distinctive ability to deliver the right product to the right place at the right time. This idea of course has become classical in economics.

Next, with regard to inputs or imports, all may say that the tolerable variation in the 'input' markets is likewise dependent upon the technological component. Thus some enterprises are enabled by their particular technical organization to tolerate considerable variation in the type and amount of labour they can recruit. Others can tolerate little.

Two significant features of this state of affairs from the point of view of open system theory may be stated as follows.

The first point is this: that there is no simple one-to-one relation between variations in inputs and outputs. Different combinations of inputs may be handled to yield similar outputs and different 'product mixes' may be produced from similar inputs. An enterprise will tend to react in this way rather than make structural changes in its organization. One of the additional characteristics of 'open systems' is that while they are in constant commerce with their environment they are selective and, within limits, self-regulating.

The second, and no less important point, is that the technological component, in converting inputs into outputs, plays a major role in determining this self-regulating property. It functions as one of the major boundary conditions of the social system in mediating between the objectives of the enterprise and its external environment. Because of this the materials, machines and territory that go to making up the technological component are usually defined, in any modern society, as "belonging" to the enterprise and excluded from similar control by other enterprises. They represent, as it were, an 'internalized environment'.

Thus these—always, of course, from the social point of view—mediating boundary conditions of this internalized environment must be represented amongst "the open system constants" which, as von Bertelanffy suggests, define the conditions under which a steady state can be achieved. As the technological component plays a key mediating role, it follows that the open system concept must be referred to the *socio-technical system*, not simply to the social system of an enterprise.

Study of a productive system therefore as an operating entity requires detailed attention to both the technological and the social components themselves both treated as systems. It is not possible to understand what is going on simply in terms of some arbitrarily selected single aspect of the technology such as the repetitive nature of the work, the coerciveness of the assembly conveyor or the piecemeal nature of the task. However, this is what is too often attempted. In the extreme case, of course, the technological component is entirely neglected. As Peter Drucker observes:—

> It has been fashionable of late, particularly in the "human relations" school, to assume that the actual job, its technology, and its mechanical and physical requirements are relatively unimportant compared to the social and psychological situation of men at work.

This is the inevitable end result of approaching the enterprise purely as a social system—and a closed system at that.

Even when there has been a detailed study of the technology this has not been systematically related to the social system but has been treated as background information.

In the earliest Tavistock study of production systems in coal mining it became apparent that so close was the relationship between these two aspects that the social and the psychological could be understood only in terms of the detailed engineering facts and of the way in which the technological system as a whole behaved in the environment of the underground situation. We broke therefore with the earlier tradition of social research in this field and embarked on a systematic attempt to elucidate the relations between the social and technological systems, each taken as wholes, whatever the level at which the study was being made. We also related this level to at least the adjacent levels, above and below.

Though an analysis of a technological system in these terms can produce a systematic picture of the tasks and task interrelations required by a technological system, between these requirements and the social system there is not a strictly determined one-to-one relation but rather what is referred to as a correlative relation.

In a very simple operation such as manually moving and stacking railway sleepers ('ties') there may well be only a single suitable work relationship structure, namely, a co-operating pair with each man taking an end of the sleeper and lifting, supporting, walking and throwing in close co-ordination with the other man. The ordinary production process is, however, much more complex and there it is unusual to find that only one particular work relationship structure can be fitted to these tasks.

This element of choice, and the mutual influence of technology and social system, I will now illustrate from some of our more recent studies made over several years, of work organization in deep seam coal mining. The data which I have circulated are adapted from an unpublished monograph by Murray and myself.

Table 1 indicates the main features of two very different forms of organization that have both been operated economically within the same seam and with identical technology. [See Appendix at end of article for details of the composite mining system.]

The conventional system combines a complex formal structure with simple work roles: the composite system combines a simple formal structure with complex work roles. In the former the miner has a commitment to only a single part task and enters into only a very limited number of unvarying social relations that are sharply divided between those within his particular task group and those who are outside. With those 'outside' he shares no sense of belongingness and he recognizes no responsibility to them for the consequences of his actions. In the composite system the miner has a commitment to the whole group task and consequently finds himself drawn into a variety of tasks in co-operation with different members of the total group; he may be drawn into any task on the coal-face with any member of the total group and do his share on any shift.

That two such contrasting social systems can effectively operate the same technology is clear enough evidence that there exists an element of choice in designing a work organization.

However, it is far from a matter of indifference which form of organization is selected. As has already been stated, the technological system and the effectiveness of the total production system will depend upon the adequacy with which the social system is able to cope with these requirements. Although alternative social systems may survive in that they are both accepted as "good enough" this does not preclude the possibility that they may differ in effectiveness.

Once the fact that there are alternatives is grasped, the question

Table 1. Same Technology, Same Coalseam, Different Social Systems

	A Conventional Cutting Long-wall Mining System.	A Composite Cutting Long-wall Mining System.
Number of men	41	41
No. of completely segregated task groups	14	1
Mean job variation for members:		
—task groups worked with	1.0	5.5
—main tasks worked	1.0	3.6
—different shifts worked	2.0	2.9

naturally arises of which will provide the optimum conditions as distinct from those which are just good enough. The design and development of optimum socio-technical systems becomes therefore a field which is now open to systematic study by combined teams of engineers and social scientists.

In the present case the composite systems consistently showed a superiority over the conventional in terms of production and costs.

This superiority reflects, in the first instance, the more adequate coping in the composite system with the task requirements. The constantly changing underground conditions require that the already complex sequence of mining tasks undergo frequent changes in the relative magnitudes and even the order of these tasks. These conditions optimally require the internal flexibility possessed in varying degrees by the composite systems. It is difficult to meet variable task requirements with any organization built on a rigid division of labour. The only justification for a rigid division of labour is a technology which demands specialized non-substitute skills and which is, moreover, sufficiently superior, as a technology, to offset the losses due to rigidity. The conventional longwall cutting system has no such technical superiority over the composite to offset its relative rigidity—its characteristic inability to cope with changing conditions other than by increasing the stress placed on its members, sacrificing smooth cycle progress or drawing heavily upon the negligible labour reserves of the pit.

The superiority of the composite system does not, however, rest alone in more adequate coping with the tasks. It also makes better provision to the personal requirements of the miners. Mutually supportive relations between task groups are the exception in the conventional system and the rule in the composite. In consequence, the conventional miner more frequently finds himself without support from his fellows when the strain or size of his task requires it. Crises are more likely to set him against his fellows and hence worsen the situation.

Table 2. Production and Costs for Different Forms of Work Organization with Same Technology

	Conventional	Composite
Productive achievement*	78	95
Ancillary work at face (hrs. per man-shift)	1.32	0.03
Average reinforcement of labour (percent of total face force)	6	—
Percent of shifts with cycle lag	69	5
No. consecutive weeks without losing a cycle	12	65

*Average percent of coal won from each daily cut, corrected for differences in seam transport.

Similarly, the distribution of rewards and statuses in the conventional system reflects the relative bargaining power of different roles and task groups as much as any true differences in skill and effort. Under these conditions of disparity between effort and reward any demands for increased effort are likely to create undue stress. The undue stress created by conventional longwall conditions is in my view the major cause of the high absence rates commonly found among face-workers. Halliday, working on data collected in Scotland before the war, found the incidence of stress illnesses to be 2½ times greater among miners than in any other occupational group in the insured population.

Table 3 indicates the difference in stress experienced by miners in the two systems. In a separate study made by Hill and myself in a steelworks, it was shown that once the role of stayer as distinct from leaver is accepted, the effects of work stress are reflected in temporary withdrawal of various kinds from the work situation and that all forms of absence are positively correlated. This gives the rationale for summing such figures in the present case.

I should like to pursue this question of stress effect a little further by commenting for a few moments on a special study carried out by Murray. . . .

The findings in mining were replicated by experimental studies in textile mills in the radically different setting of Ahmedabad, India. In this case, the social scientist—my colleague A. K. Rice—had a role which enabled him to play a leading part with those concerned in designing the new socio-technical system. The idea I mentioned a little while back of system design teams including a social scientist is therefore already more than a mere aspiration.

However, two possible sources of misunderstanding need to be considered:

1. Our findings do not suggest that work group autonomy should be maximized in all productive settings. There is an optimum level of

Table 3. Stress Indices for Different Social Systems

	Conventional	Composite
Absenteeism		
(Percent of possible shifts)		
Without reason	4.3	0.4
Sickness or other	8.9	4.6
Accidents	6.8	3.2
Total:	20.0	8.2

grouping which can be determined only by analysis of the requirements of the technological system. Neither does there appear to be any simple relation between level of mechanization and level of grouping. In one mining study we found that in moving from a hand-filling to a machine-filling technology, the appropriate organization shifted from an undifferentiated composite system to one based on a number of partially segregated task groups with more stable differences in internal statuses.

2. Nor does it appear that the basic psychological needs being met by grouping are workers' needs for friendship on the job, as is frequently postulated by advocates of better 'human relations' in industry. Grouping produces its main psychological effects when it leads to a system of work such that the workers are primarily related to each other by way of the requirements of task performance and task interdependence. When this task orientation is established the worker should find that he has an adequate range of mutually supportive roles (mutually supportive with respect to performance and to carrying stress that arises from the task). As the role system becomes more mature and integrated, it becomes easier for a worker to understand and appreciate his relation to the group. Thus in the comparison of different composite mining groups it was found that the differences in productivity and in coping with stress were not primarily related to differences in the level of friendship in the groups. The critical prerequisites for a composite system are an adequate supply of the required special skills among members of the group and conditions for developing an appropriate system of roles. Where these prerequisites have not been fully met, the composite system has broken down or established itself at a less than optimum level. The development of friendship and particularly of mutual respect occurs in the composite systems but the friendship tends to be limited by the requirements of the system and not to assume unlimited disruptive forms such as were observed in conventional systems and have been reported by Adams to occur in certain types of bomber crews.

The textile studies yielded the additional finding that *supervisory roles* are best designed on the basis of the same type of socio-technical analysis. It is not enough simply to allocate to the supervisor a list of responsibilities for specific tasks and perhaps insist upon a particular style of handling men. The supervisory roles arise from the need to control and co-ordinate an incomplete system of men-task relations. Supervisory responsibility for the specific parts of such a system is not easily reconcilable with responsibility for overall aspects. The supervisor who continually intervenes to do some part of the productive work may be proving his willingness to work, but is also likely to be neglecting his main task of controlling and co-ordinating the system so that the operators are able to get on with their jobs with the least possible disturbance.

Definition of a supervisory role presupposes analysis of the system's requirements for control and co-ordination and provision of conditions that will enable the supervisor readily to perceive what is needed of him and to take appropriate measures. As his control will in large measure rest on his control of the boundary conditions—those activities relating to a larger system—it will be desirable to create 'unified commands' so that the boundary conditions will be correspondingly easy to detect and manage. If the unified commands correspond to natural task groupings, it will also be possible to maximize the autonomous responsibility of the work group for internal control and co-ordination, thus freeing the supervisor for his primary task. A graphic illustration of the differences in a supervisory role following a socio-technical re-organization of an automatic loom shed can be seen in the two figures attached: Figure 1

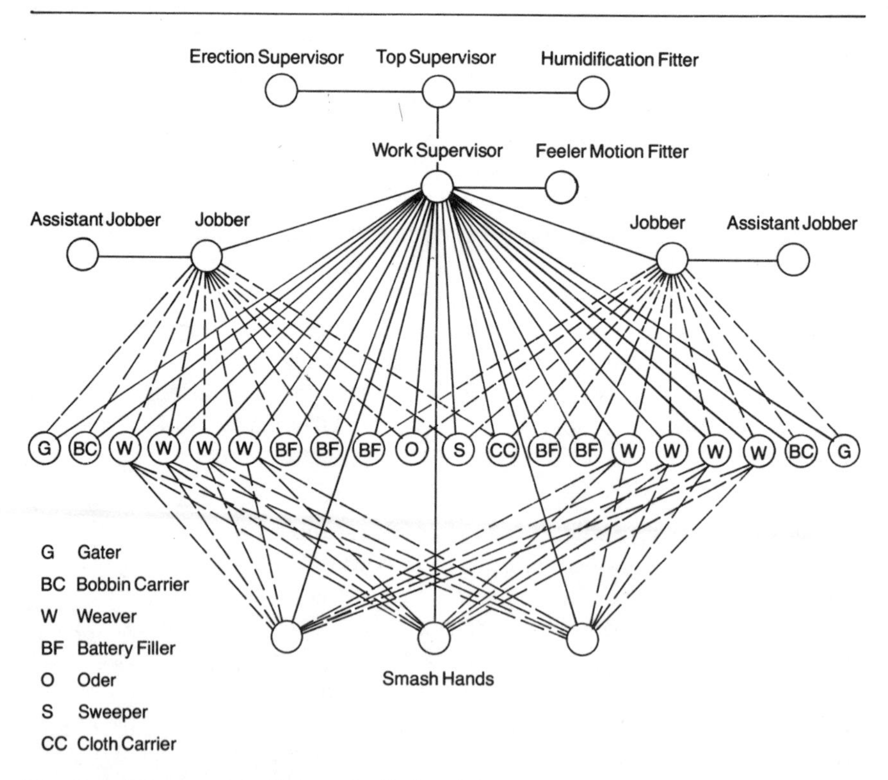

Figure 1. Management Hierarchy Before Change

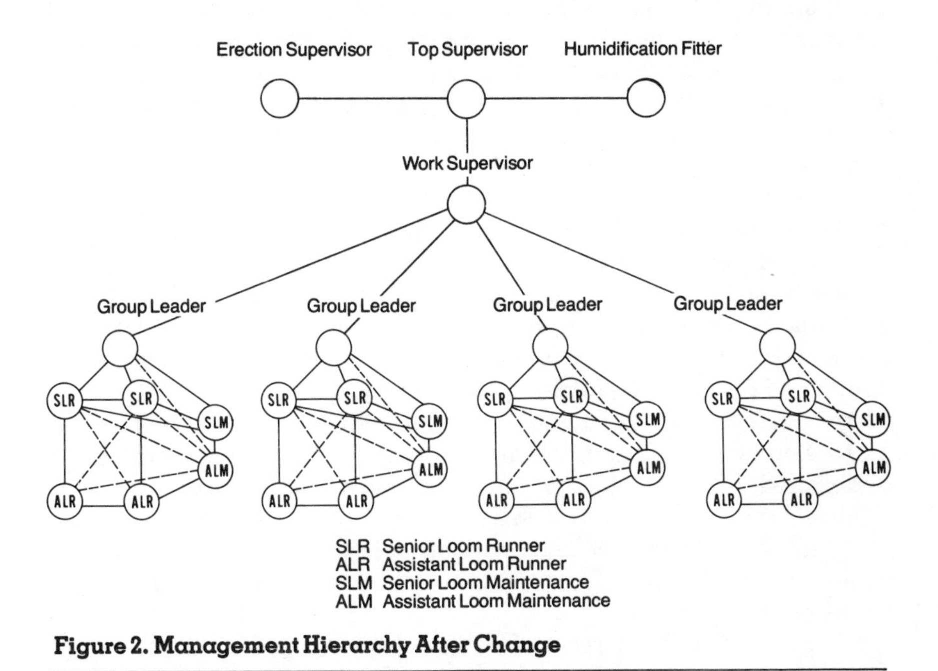

Figure 2. Management Hierarchy After Change

representing the situation before, and Figure 2 representing the situation after change.

This re-organization was reflected in a significant and sustained improvement in mean percentage efficiency and a decrease in mean percentage damage. . . . After certain setbacks which can be accounted for, the improvements were consistently maintained throughout a long period of follow-up.

The significance of the difference between the two organizational diagrams does not rest only in the relative simplicity of the latter (although this does reflect less confusion of responsibilities) but also in the emergence of clearly distinct areas of command which contain within themselves a relatively independent set of work roles together with the skills necessary to govern their task boundaries. In like manner the induction and training of new members was recognized as a boundary condition for the entire shed and located directly under shed management instead of being scattered throughout subordinate commands. Whereas the former organization had been maintained in a steady state only by the constant and arduous efforts of management, the new one proved to be inherently stable and self-correcting, and consequently freed management to give more time to their primary task and also to introduce a third shift, hitherto strongly resisted.

Similarly, the primary task in managing the enterprise as a whole is to relate the total system to its environment and is not in internal regulation *per se*. This does not mean that managers will not be involved in internal problems but that such involvement will be oriented consciously or unconsciously to certain assumptions about the external relations of the enterprise.

This contrasts with the common postulate of closed system structural-functional theories—namely that "the basic need of all empirical systems is the maintenance of the integrity and continuity of the system itself". It contrasts also with an important implication of this postulate, namely that the primary task of management is "continuous attention to the possibilities of encroachment and to the forestalling of threatened aggression or deleterious consequences from the actions of others". In industry this Maginot Line mentality represents the special and limiting case of a management that takes for granted a previously established definition of its primary task. Such managers assume that all they have to do, or can do, is sit tight and defend their market position. This is, however, the common case in statutorily established bodies and it is on such bodies that recent studies of bureaucracy have been largely carried out.

In general the leadership of an enterprise must be willing to break down an old integrity—even to create profound discontinuity—if such steps are required to take advantage of changes in technology and markets. The very survival of an enterprise may be threatened by its inability to face up to such demands, as for instance, switching the main effort from production of processed goods to marketing, or from production of heavy industrial goods to consumer goods. Similarly, the leadership may need to pay 'continuous' attention to the possibilities of making their own encroachments rather than be obsessed with the possible encroachment of others—on the Maginot principle.

Considering enterprises as 'open socio-technical systems' helps to provide a more realistic picture of how they are both influenced by and able to act back on their environment. It is a frame of reference which points in particular to the various ways in which enterprises are enabled by their structural and functional characteristics ('system constants') to cope with the 'lacks' and 'gluts' in their available environment. Unlike mechanical and other inanimate systems, they possess the property of 'equi-finality'; they may achieve a steady state from differing initial conditions and in differing ways. Thus in coping with internal change they are not limited to simple quantitative change and increased uniformity but may, and usually do, elaborate new structures and take on new functions. The cumulative effect of coping mainly by *internal* elaboration and differentiation is generally to make the system independent of an increasing range of the predictable fluctuations in its supplies and outlets. At the same time, however, this process ties down in specific ways more and more of its capital, skill and energies and renders it less able to cope with newly emergent and unpredicted change that challenge the primary ends of the enterprise. This process has been traced out in a great many empirical studies of bureaucracies.

However, there are available to an enterprise other aggressive strategies that seek to achieve a steady state by transforming the environment. Thus an enterprise has possibilities of moving into new markets or inducing changes in the old; of choosing differently than it has done from among the range of personnel, resources and technologies offered by its environment or of training and developing cadres and equipment. It can develop new consumer needs or stimulate old ones.

Thus, arising from the nature of the enterprise as an open system, management is concerned with 'managing' both an internal system and an external environment. To regard an enterprise as a closed system and concentrate upon management of the 'internal enterprise' is to expose the enterprise to the full impact of the vagaries of the environment. In several of the client organizations with which the Tavistock Institute has been concerned, we have encountered disastrous instances of this—some of them in firms which have made a specious hobby of good human relations as a defense against facing up to some of the more awkward problems of the business.

If management is to control internal growth and development, it must in the first instance control the 'boundary conditions'—the forms of exchange between the enterprise and its environment. As we have seen, most enterprises are confronted with a multitude of actual and possible exchanges. If resources are not to be dissipated the management must select from the alternatives a course of action. The casual texture of competitive environment is such that it is extremely difficult to survive on a simple strategy of selecting the best from among the alternatives immediately offering. Some that offer immediate gain lead nowhere, others lead to greater loss; some alternatives that offer loss are avoidable, others are unavoidable, if long run gains are to be made. The relative size of the immediate loss or gain is no sure guide as to what follows. Since also the actions of an enterprise can improve the alternatives that are presented to it, the optimum course is more likely to rest in selecting a strategic objective to be achieved in the long run. The optimum strategic objective is one that will place the enterprise in a position in its environment where it has assured conditions for growth. Achieving this position is the *primary task* or overriding mission of the enterprise.

In selecting an overriding primary task of this kind, it needs to be borne in mind that the relations with the environment may vary with (a) the productive efforts of the enterprise in meeting environmental requirements; (b) changes in the environment that may be induced by the enterprise, and (c) changes independently taking place in the environment. These will be of differing importance for different enterprises and for the same enterprise at different times. Managerial control will usually be greatest if the primary task can be based on productive activity. If this is not possible, as in commerce, the primary task will give more control if it is based on marketing than simply on fore-knowledge of the independent environmental changes. Managerial control will be further enhanced if the primary task, at whatever level it is selected, is such as to enable the enterprise to achieve *vis-a-vis* its competitors, a *distinctive competence*. Conversely, in our experience, an enterprise which has long occupied a favoured position because of distinctive productive competence may have grave difficulty in recognizing when it is losing control owing to environmental changes beyond its control. Anyone in consulting industrial practice is only too familiar with instances of this kind.

As Selznick has pointed out, an appropriately defined primary task offers stability and direction to an enterprise, protecting it from adventurism on the one hand or costly drifting on the other. These advantages, however, as he illustrates, may be no more than potential unless the top management group of the organization achieves solidarity about the new primary task. If the vision of the task is locked up in a single man or is the subject of dissension in top management, it will be subject to great risk of distortion and susceptible to violent fluctuations. Similarly, the enterprise as a whole needs to be reoriented and reintegrated about this primary task. Thus, if the primary task shifts from heavy industrial goods to durable consumer goods, it would be necessary to ensure that there is a corresponding shift in values that are embodied in such sections as the sales force and design department.

A theory of enterprises as open socio-technical systems must be prepared to look at these wide problems of top management just as much as at the more precise and restricted problems concerning primary work groups and departmental organizations. Though the bulk of this paper has been concerned with these narrower problems—because it is on these that most data is to hand—this survey would have been incomplete without at least a glimpse of the wider areas. It is into these areas especially, however, that social scientists must now seek to enter. From what experience I have so far had of them, I would say it is very difficult to get admitted, but, even more, once there, to understand what is going on.

APPENDIX

Characteristics of Composite System in the Seam

Work Method:	The oncoming shift takes up the cycle at whatever point the previous shift left it and carries on with whatever jobs have next to be done.
Workmen:	Multi-skilled miners, all qualified in filling, pulling, stonework and often also in drilling and cutting.
Work Groups:	Self-selected autonomous teams responsible for allocating themselves to the various jobs that management requires them to fill. Systems for the rotation of tasks and shifts among members used to regulate deployment.
Method of Payment:	An 'all-in' flat rate plus a piece-rate bonus determined by cubic yards of coal produced. Common pay note divided equally among team.

Differences Between Panels in Face Group Organization

NO. 1 PANEL	NO. 2 PANEL
'Face-wide'. Organized as two separate face teams.	'Panel-wide'. Organized as two main alternating shift groups over the whole panel.
'One-task' jobs: men tend to work at only one main task.	'Multi-task' jobs: men rotate tasks systematically and over time carry out a range of them.
Each work place and task 'tied' to a particular man.	Work places and tasks not tied to particular individuals.
Not customary for men to move from one work group to another.	Men move freely from one work group to another.

A New Role for the Work Group

Maxine Bucklow

Membership in the small primary work group has been considered a major source of motivation for employees since the Hawthorne studies. Much of the early evidence from research in group dynamics and from T-group training supported this assumption. Disappointing results from long-term studies within organizations have, however, modified early enthusiasm. This has led to current concern with problems of power equalization and the role of the rank-and-file worker. Attention is drawn to recent work of the Tavistock Institute of Human Relations with autonomous work groups, which goes far towards solving the problems of worker motivation, participation, and power equalization.

In the most recent reporting of their work, Trist and his colleagues have reformulated their theoretical position.[1] The concept that integrates the technological, economic, and sociopsychological aspects of a production system is the primary task: the work it has to perform. Work is the key transaction which relates an operating group to its environment and allows it to maintain a steady state. The concept of organizational choice is introduced so as to direct attention to the existence of a range of possible production systems. The task of management is to choose that which best fits the technical and the human requirements.

Major theoretical importance is now given to the concept of responsible autonomy. The organization of small autonomous work groups has

---

Extracted from Maxine Bucklow, "A New Role for the Work Group." *Administrative Science Quarterly*, June 1966, *11* (1), 72-74. Used with permission.

[1]E. L. Trist, G. W. Higgin, H. Murray, and A. B. Pollock, *Organizational Choice* (London: Tavistock, 1963).

been demonstrated in mining and textile situations. Success with composite longwall groups of forty men would widen the practical implications of the concept.

Responsible autonomy is seen as crucial for the satisfactory design of production systems. It gives the work group a central role in the production system, not the peripheral supporting role envisaged by Mayo and Likert, and has successfully motivated rank-and-file workers to greater cooperative effort than other methods. It also makes more basic changes in the distribution of control and power, by transferring some of the traditional authority of management for the control and coordination of jobs, i.e., the part appropriate to the primary group's task, to the men who actually perform the task. Trist criticizes the proposals of McGregor and Likert to achieve these ends, for failing to understand the difficulties involved, particularly the initial anxiety at relaxing traditional management controls over the primary group.

This real transferring of power and control to the group for the operation of the primary task has other advantages. The coal study supports other evidence that increasing control at lower levels does not decrease control at higher levels nor adversely affect efficiency. As Trist suggests, it exerts an upwards pressure in the managing system which affects all roles, so that all levels have more, rather than less, opportunity to carry out their managerial roles in a broader way. Trist now believes that the transfer of some control to autonomous work groups is the only means of overcoming the split at the bottom of the executive system at Glacier Metal Co.

Emery has recently reassessed the Tavistock work at Glacier, and criticizes the early concern with the working through of problems and with the formal aspects of industrial democracy, without making any basic change in the role of the rank-and-file worker. He now sees the development of autonomous work groups as "the democratisation of the work place" and suggests that industrial democracy, while making decisions more democratic, has not altered the content of a worker's relation to his job.[2]

Herbst who made the first detailed day-to-day study of the interactions of a composite group of miners, criticized the Morse and Reimer study for changing only the locus of decision making and not the activities about which decisions were made. He suggested that joint participation in the task may be a necessary prerequisite for joint decision making to be maintained.[3]

---

[2]F. E. Emery, Technology and Social Organization, *Scientific Business*, 1(1963), 132-136.

[3]P. G. Herbst, *Autonomous Group Functioning* (London: Tavistock, 1962).

It has been argued that the Tavistock concept of the autonomous work group goes far towards solving some of the problems of worker motivation, participation, and power equalization, with which American researchers are preoccupied. The Tavistock concept also provides a new role for the work group different from that advocated by Mayo, Lewin, and Likert. The reorganized groups at Non-Linear Systems, which were virtually autonomous, give further support to the Tavistock concept. King's reorganizations and retraining of women in a Norwegian clothing factory can also be cited as supporting evidence. They were given responsibility for control over their work and work organization, and the result was an increase in output and satisfaction and a broadening of the functions of the unit manager.[4]

---

[4]D. King, *Training Within the Organization* (London: Tavistock, 1964).

The Principles of Sociotechnical Design

Albert Cherns

The art of organization design is simultaneously esoteric and poorly developed. Most existing organizations were not born but "just growed." Many bear the recognizable stigmata of the operations of various well-known consultancy groups. There is, of course, no lack of available models and no one seeking to set up an organization need invent the wheel. But organization design is generally an outcome not an input. The input in manufacturing organizations is provided by the engineers, both those who design machines and equipment and those who design work methods and layout—the industrial engineers. Increasingly, operations researchers, systems analysts, the designers of computerized information systems, and the providers of "management services" of all kinds are having their say. In nonmanufacturing work organizations, it is the latter who are most influential. And all of them, whether they recognize it or not, bring assumptions about people into their operations and their design. Most simply put, these assumptions can generally be described as Taylorist or System X. People are unpredictable. If they are not stopped by the system design, they will screw things up. It would be best to eliminate them completely; but since this is not possible, we must anticipate all the eventualities and then program

---

I am indebted to Louis E. Davis, on whose work in designing new organizations I have drawn heavily in this article, which arises out of the courses we have given together at UCLA and elsewhere.

Reprinted from Albert Cherns, "The Principles of Sociotechnical Design," HUMAN RELATIONS, 1976, *29* (8), 783-792. Used with permission of Plenum Publishing Corporation.

them into the machines. The outcome is the familiar pattern of hierarchies of supervision and control to make sure that people do what is required of them, and departments of specialists to inject the "expert" knowledge that may be required by the complexities of manufacturing, marketing, and allied processes, but is equally often required to make the elaborate control, measurement, and information systems work.

We have found in our own work, both teaching and consulting, that engineers readily perceive that they are involved in organization design and that what they are designing is a sociotechnical system built around much knowledge and thought on the technical and little on the social side of the system. There is, of course, the danger that the term *sociotechnical system* very rapidly becomes a shibboleth, the mere pronouncing of which distinguishes the *cognoscenti* from the ignorant and uninitiated. But recognizing that a production system requires a social system to integrate the activities of the people who operate, maintain, and renew it; account for it; and keep it fed with the resources it requires and dispose of the products does nothing by itself to improve the design. And while discussion of the characteristics of social systems is helpful, that still leaves us with the problem that there are many ways of achieving their essential objectives.

We teach engineers that any social system must, if it is to survive, perform the function of Parson's (1951) four subsystems. As we present them, these functions are attainment of the goals of the organization; adaptation to the environment; integration of the activities of the people in the organization, including the resolution of conflict whether task-based, organization-based, or interpersonally based; and providing for the continued occupation of the essential roles through recruitment and socialization. The advantage of this analysis is that it tells the designer that if he doesn't take these absolute requirements of a social system into account, he will find that they will be met in some way or other, quite probably in ways that will do as much to thwart as to facilitate the functions for which he does plan. But it still leaves wide open the question of how to design a social system or, more fundamentally, how much a social system should be designed. That there is a choice in such matters can be as much a revelation to the engineer as the fact that there is a choice of technology to achieve production objectives is to the social scientist.

How, then, do you design a sociotechnical system? Can we communicate any principles of sociotechnical design? The first thing to be said is that a lot depends upon your objectives. As we have said, all organizations are sociotechnical systems; that is no more than a definition, a tautology. But the phrase was first used with, and has acquired, the connotation that organizational objectives are best met not by the optimization of the technical system and the adaptation of a social system to it, but by the joint optimization of the technical and the social aspects, thus exploiting the adaptability and innovativeness of people in attaining goals instead of overdetermining technically the manner in which these goals should be attained.

It is an obvious corollary that such design requires knowledge of the way machines and technical systems behave and of the way people and groups behave. Unless a designer is himself an engineer and a social scientist, both are required, which means engineers discussing alternative technical ways of attaining objectives with social scientists. This is not easy unless social scientists will take the trouble to learn enough about technology to understand the kinds of options that are open to engineers. The design team has indeed to be a multifunctional one as we have described elsewhere (Cherns, 1972).

In the process of designing ideas, no doubt the constant interchange among engineer, manager, social scientist, financial controller, personnel specialist, and so on can do much to ensure that all aspects are considered, but the sociotechnical concepts involved need not be hammered out afresh every time. They can be collected and presented in such a way as to ensure that they are taken account of, yet not straitjacketing the designer. To this end, we have described nine principles, which we offer as a checklist, not a blueprint. They represent a distillation of experience and owe more to the writings of others (Emery & Trist, 1972; Herbst, 1974) than to our own originality. They have not, however, previously been systematized. These principles are:

PRINCIPLE 1: COMPATIBILITY

The process of design must be compatible with its objectives. A camel has been defined as a horse designed by a committee, and that joke unkindly incorporates negative evaluations of camels and committees. Camels certainly have minds of their own, but perhaps any attempt to draw more parallels between a camel and a social system would be unduly fanciful. Would a horse be more acceptable to a despot and a camel to a democrat? The point to be made, however, is that a participative social system cannot be created by fiat.

If the objective of design is a system capable of self-modification, of adapting to change, and of making the most use of the creative capacities of the individual, then a constructively participative organization is needed. A necessary condition for this to occur is that people are given the opportunity to participate in the design of the jobs they are to perform. In a redesign of an existing organization, the people are already there; a new design has, however, to be undertaken before most of the people are hired. To some extent their jobs will have been designed for them in advance, but this extent can be kept to a minimum. In one case (Davis & Cherns, in press) the design team took the view that they would not design other people's lives. Having defined what were to be the objectives to be met and the competences required to meet them, they deferred until the individual was appointed any discussion of how the job was to be performed. And, as in most cases, "job" was not defined, this meant involving the people appointed as a team. Clearly some decisions had and have to be taken in advance; there has to be a pretty firm notion of how many people will be required and of what kinds of competence must be sought, but this is governed by the second principle.

PRINCIPLE 2: MINIMAL CRITICAL SPECIFICATION

This principle has two aspects, negative and positive. The negative simply states that no more should be specified than is absolutely essential; the positive requires that we identify what is essential. It is of wide application and implies the minimal critical specification of tasks, the minimal critical allocation of tasks to jobs or of jobs to roles, and the specification of objectives with minimal critical specification of methods of obtaining them. While it may be necessary to be quite precise about what has to be done, it is rarely necessary to be precise about how it is to be done. In most organizations there is far too much specificity about how and indeed about what. Any careful observer of people in their work situation will learn how people contrive to get the job done in despite of the rules. As the railwaymen in Britain have demonstrated, the whole system can be brought to a grinding halt by "working to rule." Many of the rules are there to provide protection when things go wrong for the man who imposed them; strictly applied, they totally inhibit adaptation or even effective action.

In any case, it is a mistake to specify more than is needed because by doing so options are closed that could be kept open. This premature closing of options is a pervasive fault in design; it arises, not only because of the desire to reduce uncertainty, but also because it helps the designer to get his own way. We measure our success and effectiveness less by the quality of the ultimate design than by the quantity of our ideas and preferences that have been incorporated into it.

One way of dealing with the cavalier treatment of options is to challenge each design decision and demand that alternatives always be offered. This may result in claims that the design process is being expensively delayed. Design proposals may also be defended on the ground that any other choice will run up against some obstacle, such as a company practice, or a trade union agreement, or a manning problem. These obstacles can then be regarded and logged as constraints upon a better sociotechnical solution. When they have all been logged, each can be examined to estimate the cost of removing it. The cost may sometimes be prohibitive, but frequently turns out to be less formidable than supposed or than the engineer has presented it to be.

PRINCIPLE 3: THE SOCIOTECHNICAL CRITERION

This principle states that variances, if they cannot be eliminated, must be controlled as near to their point of origin as possible. We need here to define *variance*, a word much used in sociotechnical literature. Variance is any unprogrammed event; a key variance is one which critically affects outcome. This might be a deviation in quality of raw material, the failure to take action at a critical time, a machine failure, and so on. Much of the elaboration of supervision, inspection, and management is the effort to control variance, typically by action which does less to prevent variance than to try to correct its consequences. The most obvious example is the inspection function. Inspecting a product, the outcome of any activity, does not make right what is wrong. And if this inspection is carried out in a separate department some time after the event, the correction of the variance becomes a long loop which is a poor design for learning. The sociotechnical criterion requires that inspection be incorporated with production where possible, thus allowing people to inspect their own work and learn from their mistakes. This also reduces the number of communication links across departmental boundaries (see also Principle 5). The fewer the variances that are exported from the place where they arise, the fewer the levels of supervision and control that are required and the more "complete" the jobs of the people concerned, to whom it now becomes possible to allocate an objective and the resources necessary to attain it. Frequently what is required to attain this objective turns out to be the supply of the appropriate information as discussed (see Principle 6).

Identifying variances and determining the key variances is a process often requiring lengthy analysis, and from time to time efforts have been made to codify it. One version, known as the nine-step analysis, has been developed by Davis and Cherns (in press). It has been used in enough organizations to give us some assurance that it can be adapted to use with any type of work organization, not just with manufacturing industry.

PRINCIPLE 4: THE MULTIFUNCTIONAL PRINCIPLE —ORGANISM VS. MECHANISM

The traditional form of organization relies very heavily on the redundancy of parts. It requires people to perform highly specialized, fractionated tasks. There is often a rapid turnover of such people but they are comparatively easily replaced. Each is treated as a replacement part. Simple mechanisms are constructed on the same principle. Disadvantages arise when a range of responses is required, that is, when a large repertoire of performances is required from the mechanism or the organization. This usually occurs if the environmental demands vary. It then becomes more adaptive and less wasteful for each element to possess more than one function. The same function can be performed in different ways by using different combinations of elements. There are several routes to the same goal—the principle sometimes described as equifinality. Complex organisms have all gone this route of development. The computer, for example, is a typical multifunctional mechanism. The principle of minimal critical specification permits the organization to adopt this principle also.

PRINCIPLE 5: BOUNDARY LOCATION

In any organization, departmental boundaries have to be drawn somewhere. Miller (1959) has shown that such boundaries are usually drawn so as to group people and activities on the basis of one or more of three criteria: technology, territory, time. Grouping by technology is typically seen in machine shops, where all the grinding machines are in one room, the Grinding Department, the milling machines in another, the Milling Department, and so on, with each department under the supervision of a specialist, a foreman grinder, etc. The consequences of this for the scheduling of work has been well described by Williamson (1972). A part in construction may spend months shuffling between departments, spending 1% of that time actually in contact with the machines. The consequent excessive cost of such work has been one of the stimuli to "group technology," the establishment of departments which each contain a variety of machines so that a part can be completed within one department. This corresponds to a grouping on the basis of time—the contiguity in time of operations indicates that they may well be organized together. Group technology also has consequences for the operation of the department as a team with its members taking responsibility for the scheduling of operations and possibly the rotation of jobs.

Other examples of grouping on the basis of technology, but not of course group technology, are the typing pool and the telephone switchboard.

The switchboard may also be an example of the criterion of territory. Switchboard operators are bound together by the design of the machine. But the territorial principle can operate on the basis of little other than spatial contiguity. If the engineers have for convenience located different activities in the same area, the maintenance of control over the people working there suggests that they be made answerable to the same supervision. Retail trade organization is often of this kind with a floor supervisor. Organizations of this kind give rise to "dotted-line" relationships of functional responsibility.

All these criteria are pragmatic and defensible up to a point. But they possess notable disadvantages. They tend to erect boundaries which interfere with the desirable sharing of knowledge and experience. A simple example may suffice. In an organization concerned with the distribution of petroleum products studied by Cherns and Taylor (unpublished data), the clerks who collected customers' orders were organized in a department separate from that of the drivers for whom schedules were worked out. A driver would pick up a schedule allocating him a vehicle and a route. Frequently the receipt of the routing would stimulate a string of expletives from the driver: "If I do what this *** has told me to, I should not be able to do half the job. I would arrive at customer B just after 12 o'clock when the only man with the key to the pumps has gone off to his lunch break. And it's no use my turning up to customer P until I have discharged enough of my load for his short pipe to reach my tank. And finally I would end up on the *** road just in the middle of the rush hour. It would serve him right if I followed these instructions; I would run out of time [exceed the permitted number of consecutive driving hours] right in the middle of the throughway." There was no doubt pardonable exaggeration in all this; the point is that the drivers had acquired a great deal of knowledge about customers, routes, etc., but being organized into a separate department they shared very little of this knowledge with the routing clerks who, however, received the customers' complaints before the drivers.

The principle has certain corollaries. One very important one concerns the management of the boundaries between department and department, between department and the organization as a whole, and between the organization and the outside world. The more the control of activities within the department becomes the responsibility of the members, the more the role of the supervisor/foreman/manager is concentrated on the boundary activities—ensuring that the team has adequate resources to carry out its functions, coordinating activities with those of other departments, and foreseeing the changes likely to impinge upon them. This boundary maintenance role is precisely the requirement of the supervisor in a well-designed system.

Under favorable circumstances, working groups can acquire and handle a greater degree of autonomy and learn to manage their own boundaries. This implies locating responsibility for coordination clearly and firmly with those whose efforts require coordination if the common objectives are to be achieved. The role of supervisor now becomes that of a "resource" to the working group.

PRINCIPLE 6: INFORMATION FLOW

This principle states that information systems should be designed to provide information *in the first place* to the point where action on the basis of it will be needed. Information systems are not typically so designed. The capacity of computer-controlled systems to provide information about the state of the system both totally and in great detail to any organizational point has been used to supply to the top echelons of the organization information which is really useful only at lower levels and which acts as an incitement to the top management to intervene in the conduct of operations for which their subordinates are and should be responsible. The designer of the information system is naturally concerned to demonstrate its potentialities and is hard to convince that certain kinds of information can be potentially harmful when presented to high organizational levels. Properly directed, sophisticated information systems can, however, supply a work team with exactly the right type and amount of feedback to enable them to learn to control the variances which occur within the scope of their spheres of responsibility and competence and to anticipate events which are likely to have a bearing on their performance.

PRINCIPLE 7: SUPPORT CONGRUENCE

This principle states that the systems of social support should be designed so as to reinforce the behaviors which the organization structure is designed to elicit. If, for example, the organization is designed on the basis of group or team operation with team responsibility, a payment system incorporating individual members would be incongruent with these objectives. Not only payment systems, but systems of selection, training, conflict resolution, work measurement, performance assessment, timekeeping, leave allocation, promotion, and separation can all reinforce or contradict the behaviors which are desired. This is to say that the management philosophy should be consistent and that management's actions should be consistent with its expressed philosophy. Not infrequently a management committed to philosophies of participation simultaneously adopts systems of work measurement, for example, which are in gross contradiction. Even management as progressive and committed to the humanization of work as that of Volvo's Kalmar plant has retained a commitment to a system of payment based on MTM, a technique of work measurement utilizing time and method study. Until replaced, this may, in fact, pose an obstacle to the further humanization of work at Kalmar to which the management is committed.

PRINCIPLE 8: DESIGN AND HUMAN VALUES

This principle states that an objective of organizational design should be to provide a high quality of work. We recognize that quality is a subjective phenomenon, and that not everyone wants to have responsibility, variety, involvement, growth, etc. The objective is to provide these for those who do want them without subjecting those who don't to the tyranny of peer control. In this regard we are obliged to recognize that all desirable objectives may not be achievable simultaneously.

What constitutes human work is a matter again of subjective judgment based on certain psychological assumptions. Thorsrud (1972) has identified six characteristics of a good job which can be striven for in the design of organizations and jobs. They are as follows: (1) the need for the content of a job to be reasonably demanding of the worker in terms other than sheer endurance, and yet to provide a minimum of variety (not necessarily novelty); (2) the need to be able to learn on the job and to go on learning (again it is a question of neither too much nor too little); (3) the need for some minimal area of decision-making that the individual can call his own; (4) the need for some minimal degree of social support and recognition in the workplace; (5) the need for the individual to be able to relate what he does and what he produces to his social life; and (6) the need to feel that the job leads to some sort of desirable future (not necessarily promotion).

PRINCIPLE 9: INCOMPLETION

Design is a reiterative process. The closure of options opens new ones. At the end we are back at the beginning. The Forth Bridge, in its day an outstanding example of iron technology, required painting to fend off rust. Starting at the Midlothian end, a posse of painters no sooner reached the Fife end than the Midlothian end required painting again. Varying the image, Jewish tradition prescribes that one brick be omitted in the construction of a dwelling lest the jealousy of God's angels be excited. Disregarding the superstition, the message is acceptable. As soon as design is implemented, its consequences indicate the need for redesign. The multifunctional, multilevel, multidisciplinary team required for design is needed for its evaluation and review.

CONCLUDING REMARKS

Who is the sociotechnical designer to whom this paper is especially addressed? The analysis, preparation, and implementation of a sociotechnical design is, as we have indicated, the property of no individual or set of individuals; it belongs to the members of the organization whose working lives are being designed. Special skills and knowledge may well be and often are required and these are provided as a resource by sociotechnical consultants or action researchers.

But participation by employees in the design of their organizations may imply that they accept or show readiness to accept work roles which go beyond the agreements and constraints evolved by negotiation between management and union on their behalf. Unions are thus inevitably involved in the process, whether in a collaborative, neutral, or antagonistic role. Can they be partners in design? This is a role which has seldom been offered to, and even more rarely accepted by, unions. It is not a role for which they have prepared themselves, and it is one which could easily blur their primary responsibilities to their members. Yet without them the viability of the design must be in some doubt. And the design of a social support system implies designing the functions of the shop steward if not the union official. Our first principle, compatibility, requires that the unions be brought into the design if that is at all possible. But if they are to come in, they, too, will need to acquire new competences. Unions are organizations, but the first is yet to be the client of a sociotechnical design.

REFERENCES

CHERNS, A. B. Helping managers: What the social scientist needs to know. *Organizational Dynamics*, Winter 1973, 51-67.

DAVIS, L. E., & CHERNS, A. B. *Designing organizations around human values* (2 vols.). New York: Free Press (in press).

EMERY, F. E., & TRIST, E. L. *Towards a social ecology*. London: Plenum Press, 1972.

HERBST, P. G. *Sociotechnical design*. London: Tavistock, 1974.

MILLER, E. J. Technology, territory and time: The internal differentiation of complex production systems. *Human Relations*, 1959, *12*, 243-272.

PARSONS, T. *The social system*. London: Routledge and Kegan Paul, 1951.

THORSRUD, E. Policy making as a learning process. In A. B. Cherns, R. Sinclair, & W. I. Jenkins, (Eds.), *Social science and government: Policies and problems*. London: Tavistock, 1972.

WILLIAMSON, D. T. N. *The anachronistic factory*. Proceedings of the Royal Society, A331, 1972, 139-160.

PART IV-A:

SOCIOTECHNICAL DIAGNOSIS

Introduction

This section of the book, dealing with sociotechnical diagnosis and change, is by far the most extensive. Although the preceding theoretical articles make important contributions in their own right, the greatest utility of a theory is in its application. Contrary to what is common in some applied behavioral science disciplines, practice has for the most part preceded the theoretical development of sociotechnical concepts. The articles in this section are primarily reports on applications that have been tried and tested. Although some articles are research oriented, most are easy-to-follow blueprints for diagnosis and change.

Woodward, Miller, and Emery and Trist deal with diagnosing sociotechnical systems. Each article is a unique contribution to the analysis of organizations from a sociotechnical system perspective, with particular emphasis on data relating to potential changes.

Woodward gives the findings of her classic study of the impact of technological complexity on organizations. By collecting data from one hundred firms, Woodward was able to discern the impact of technology on such organizational variables as levels of authority, scope of control of first-line supervisors, ratio of direct to indirect labor, ratio of managers to total personnel, specialization between functions of management, organizational flexibility, and the span of control of the chief executive. The conclusion of her study was that the characteristics of successful firms in each technological category approximated the median for that category. Less successful firms tended to utilize organizational designs and features that deviated widely from the median characteristics of their particular technological groups. This finding shows that certain organizational features are dictated by the objectives of the firm and by the technology used to achieve them. This has obvious relevance for sociotechnical diagnosis and system design. Woodward also found that the technology employed by the firm is a crucial factor in determining the attitudes and behaviors of both employees and management. These "situational demands," as she calls them, need to be considered when diagnosing organizations for sociotechnical change.

Although Miller's article is not research based, it complements the other papers at a conceptual level. Miller is concerned with the internal structure of the organization, particularly the formation of subsystems. He notes that organizations become internally differentiated on three principal dimensions: technology, territory, and time. He suggests that the efficiency of a sociotechnical system is largely determined by how it handles the forces that push toward greater differentiation of the organization. Of the many possible internal divisions of organizations, Miller affirms that the most effective arrangement is one that corresponds most closely with requirements for the performance of the primary task. He notes that if the situation demands either premature or postponed differentiation it will lead to organizational inefficiencies, stress, and managerial difficulty integrating subsystem output for the achievement of goals. Using a number of industrial examples, Miller illustrates the application of his three principles of differentiation to organizational design and management. His work is particularly important to the sociotechnical analyst because the first step in sociotechnical diagnosis is to define the boundaries of the system. Although placing boundaries around internal organizational systems may be straightforward in some cases, existing dimensions cannot be accepted without question. In some cases, redefining the boundaries according to the principles of technology, territory, and time may be the only way to procure successful sociotechnical change.

Emery and Trist give an analytical model for sociotechnical systems that includes: doing an initial scanning; identifying unit operations; identifying key process variances; analyzing social systems; collecting employee perceptions of their roles; diagnosing the maintenance system; diagnosing the supply and user systems; determining organizational development plans; and making proposals for change. This model is designed primarily for use in settings associated with physical technologies. A second model, which has not been extensively tested, concentrates on organizational objectives and role analysis for service and professional settings. Both models are well documented. Additional information on them and on their application is presented in Cummings and Srivastva (1977).

REFERENCE

Cummings, T., & Srivastva, S. *The management of work: A sociotechnical approach*. Kent, Ohio: Kent State University Press, 1977.

Management and Technology

Joan Woodward

INTRODUCTION

The research described in this booklet was the first attempt in Britain to discover whether the principles of organization laid down by an expanding body of management theory correlate with business success when put into practice.

It was carried out between 1953 and 1957 by the Human Relations Research Unit of the South East Essex Technical College. The original intention of the research workers was to look at the division of responsibilities between line supervision and the technical specialists who apply technology to the production process, and at the factors which determine the relationships between them. They soon found, however, that this line-staff relationship could not be studied in isolation, so they widened their investigations to include the whole structure of management and supervision. Their basic survey in 91 per cent of the manufacturing firms in south Essex with over 100 employees revealed considerable variations in the pattern of organization which could not be related to size of firm, type of industry or business success.

When, however, the firms were grouped according to similarity of objectives and techniques of production, and classified in order of the

---

Reprinted from Joan Woodward, *Problems of Progress in Industry No. 3–Management and Technology*. Reproduced with permission of the Controller of Her Britannic Majesty's Stationery Office, London.

technical complexity of their production systems, each production system was found to be associated with a characteristic pattern of organization. It appeared that technical methods were the most important factor in determining organizational structure and in setting the tone of human relationships inside the firms. The widely accepted assumption that there are principles of management valid for all types of production systems seemed very doubtful—a conclusion with wide implications for the teaching of this subject.

After completing the survey the team studied more fully twenty firms selected along a scale of technical complexity, and made detailed case studies of three firms in which production systems were mixed or changing.

This summary covers all three stages of the research. It describes the background survey, giving enough of the information collected to show some of the main differences between the organizational patterns associated with each of the different systems of production. The more descriptive information obtained in the second stage of the research is used to provide explanations of these differences. Finally, the detailed case studies are briefly referred to and an attempt is made to show how the analysis of changes in technical demands due to innovation can help to solve in advance the problems of management organization likely to arise.

THE SURVEY

The Area

Industrial development came comparatively late to south Essex and newer industries such as oil refining, wireless, photography, pharmaceuticals, paperboard and vehicles predominate. Factory buildings are on the whole modern. So is management organization. Most factories here were built when the functions of ownership and management had already been separated and there are few long-established family businesses. A number of family firms did move here, but their history suggests that the move gave most of them an occasion for radical changes in management structure.

The Firms

The investigation was confined to manufacturing firms in the area. Those concerned with mining and quarrying, building contracting and laundering were excluded, as were transport undertakings, public utilities and local authorities.

A long search produced a list of 203 manufacturing firms which was as comprehensive as humanly possible; it is unlikely that any firm employing 100 people or more was omitted.

The number employed ranged from a dozen to approximately 35,000. (See Table I.)

There are more large firms in south Essex than in the country generally, 9 per cent employing more than 100 people as against 1.7 per cent overall. The 203 firms cover a wide range of industries. In most of them the number employed in the area represents between 1 and 2 per cent of the national total. In textiles and leather the percentage is particularly low, but in vehicles and chemicals it is as high as 7 per cent.

A 25 per cent sample survey of the 93 firms employing less than 100 people showed no clear-cut level of management between board and operators in most of them. The main survey was therefore confined to the 110 firms employing 100 people or more, of which 100, or 91 per cent, were willing and able to take part.

Of these 100 firms, 68 had both their main establishment and their commercial headquarters inside the area; the rest had only branch factories.

Information Obtained

A research worker visited each of the firms and obtained information under the following headings:

1. History, background and objectives.
2. Description of the manufacturing processes and methods.
3. Forms and procedures through which the firm was organized and operated.

Table I. Size Distribution of Manufacturing Firms in South Essex

FIRMS EMPLOYING	PERCENTAGE OF 203 FIRMS	PERCENTAGE OF LABOUR FORCE (119 400)
100 or less	46	3
101-250	24	7
251-500	12	8
501-1000	9	11
1001-2000	4	10
2001-4000	3	14
4001-8000	1	9
8000 and over	1	38
Totals	100	100

(a) An organization chart.
(b) A simple analysis of costs into three main divisions: wages, materials, and overheads.
(c) An analysis of the labour structure, including the size of the span of control at various levels and the following ratios:
  (i) Direct production workers to total personnel.
  (ii) Maintenance workers to direct production workers.
  (iii) Clerical and administrative to hourly paid personnel.
  (iv) Managers and supervisory staff to total personnel.
(d) The organization and operation of sales activities, research and development, personnel management, inspection, maintenance, and purchasing.
(e) The procedures used in production control and planning.
(f) The procedures used in cost or budgetary control.
(g) The qualifications and training of managers and supervisory staff; management recruitment and training policy.

4. Information helpful in making an assessment of the firm's efficiency.

The Assessment of Efficiency

It is not easy to assess either the success of a firm or the effectiveness of a particular administrative expedient. The circular argument that an arrangement works because it exists is difficult to avoid. But an assessment was attempted. The firms were classified into three broad categories of success: average, below average and above average. The more obvious factors considered were profitability, market standing, rate of development and future plans. Questions were asked about the unit of measure commonly applied to the product, the volume of the industry's output, the proportion of that volume produced by the firm concerned, and the nature of the market. More subjective factors considered included the reputation of the firm, both inside its industry and among local firms, the quality and attitudes of its management and supervisory staff, the rate of this staff's turnover, and the opportunity provided for a complete and satisfying career in management.

RESULTS

Organizational Differences Between Firms

The 100 firms in the survey were organized and run in widely different ways. In only about half did the principles and concepts of management theory appear to have had much influence on organizational development.

In 35 firms there was an essentially 'line' or 'military' type of organization; two firms were organized functionally, almost exactly as recommended by Taylor fifty years ago.[1] The rest followed in varying degrees a line-staff pattern of organization; that is, they employed a number of functional specialists as 'staff' to advise those in the direct line of authority.

The number of distinct levels of management between board and operators varied from two to twelve; while the span of control of the chief executive[2] ranged from two to nineteen, and that of the first line supervisor[3] from seven to ninety. (An individual's span of control is the number of people directly responsible to him.)

Wages and salaries accounted for anything between 3 per cent and 50 per cent of total costs. Labour forces differed in character from firm to firm too; for example, the ratio of clerical and administrative staff to hourly paid workers ranged between 3 : 1 and 1 : 14; and that of direct to indirect labour between 1 : 3 and 15 : 1. Exactly half of the firms employed graduates or other professionally qualified staff. Thirty firms promoted their managers entirely from within, five from outside only, and the remainder used both sources according to circumstances.

There was no obvious explanation of these differences in organizational structure; they did not appear to be related either to size or type of industry. Also, conformity with the 'rules' of management did not necessarily result in success or nonconformity in commercial failure. Of the twenty firms assessed as 'above average' in success, only nine had a clearly defined organizational pattern of the orthodox kind.

New Ideas About Management

Did any common thread underlie these differences? One possible explanation was that they reflected the different personalities of the senior managers, another that they arose from the historical background of the firms. While such factors undoubtedly influenced the situation, they did not adequately explain it; they were not always associated with differences in organizational patterns or in the quality of human relations.

---

[1]F. W. Taylor, *Shop Management*, Harper Bros., New York-London, 1910.

[2]The chief executive was in some cases the Chairman, in others the Managing Director, and in others the General or Works Manager. In every case he represented the highest level of authority operating full-time on the spot.

[3]i.e. the first level of authority that spent more than 50 percent of the time on supervisory duties.

A new approach lay in recognizing that firms differed not only in size, kind of industry and organizational structure, but also in objectives. While the firms were all manufacturing goods for sale, their detailed objectives depended on the nature of the product and the type of customer. Thus some firms were in more competitive industries than others, some were making perishable goods that could not be stored, some produced for stock, and others to orders; in fact, marketing conditions were different in every firm. The underlying purpose varied too. For example, one firm had originally undertaken manufacture to demonstrate that the products of its mines could be effective substitutes for other more commonly used materials.

These differences in objectives controlled and limited the techniques of production that could be employed. A firm whose objective was to build prototypes of electronic equipment, for example, could not employ the technical methods of mass-production engineering. The criterion of the appropriateness of an organizational structure must be the extent to which it furthers the objectives of the firm, not, as management teaching sometimes suggests, the degree to which it conforms to a prescribed pattern. There can be no one best way of organizing a business.

This is perhaps not sufficiently recognized; management theorists have tried to develop a 'science' of administration relevant to all types of production. One result is that new techniques such as operational research and the various tools of automation have been regarded as aids to management and to industrial efficiency rather than as developments which may change the very nature of management.

Evidence is accumulating, particularly in the United States, that automation and other technological changes are often associated with considerable disturbance in the management systems of the firms concerned. New tools begin to change the task and the new task begins to change the organization and the qualities required to carry it out successfully. For example, work done in the United States has shown that the qualities required of the foreman on a motor-car assembly line appear to be very different from those required on transfer-line production.[4] Expressions like 'leadership' or 'the art of foremanship', used so often in management literature, are losing much of their meaning. It is possible, for example, that leadership must be directive, participant, or *laissez-faire* according to circumstances. A good leader in one situation is not necessarily a good leader in another.

---

[4]*The Man on the Assembly Line*, Charles R. Walker and Robert H. Guest (Cambridge, Mass: Harvard University Press, 1952).

Two interesting questions have so far emerged. Are the management organization and supervisory qualities required in a firm in the process of radical technical change different from those required in a stable firm? Does the kind of organization required vary with the technical complexity of the manufacturing methods?

Differences in Technical Methods

The firms were grouped according to their technical methods. Ten different categories emerged. (See Figure 1.)

Firms in the same industry did not necessarily fall into the same group. For example, two tailoring firms of approximately equal size had very different production systems; one made bespoke suits, the other mass-produced men's clothing.

Measurement of Technical Complexity

The ten production groups listed in Figure 1 form a scale of technical complexity. (This term is used here to mean the extent to which the production process is controllable and its results predictable). For example, targets can be set more easily in a chemical plant than in even the most up-to-date mass-production engineering shops, and the factors limiting production are known more definitely so that continual productivity drives are not needed.

Some of the firms studied used techniques of operational research to increase control over production limitations—but these could be effective only within limits set by technical methods, which were always the major factor determining the extent of control over production.

Production Systems and Technical Progress

Grading firms according to their technical complexity implies no judgment of their progressiveness or backwardness, nor is it any indication of the attitude of their managements towards technical innovation. Each production system has its particular applications and limitations. While there remains a demand for the gold-plated limousine or the bespoke suit, while large items of equipment have to be built, or while progress in industries like electronics proceeds too rapidly to permit standardization, there will be a place for unit production even though it is less advanced technically than other systems. Moreover, although continuous-flow production is applicable to the manufacture of single components, it is difficult as yet to foresee its use where many different component parts are assembled.

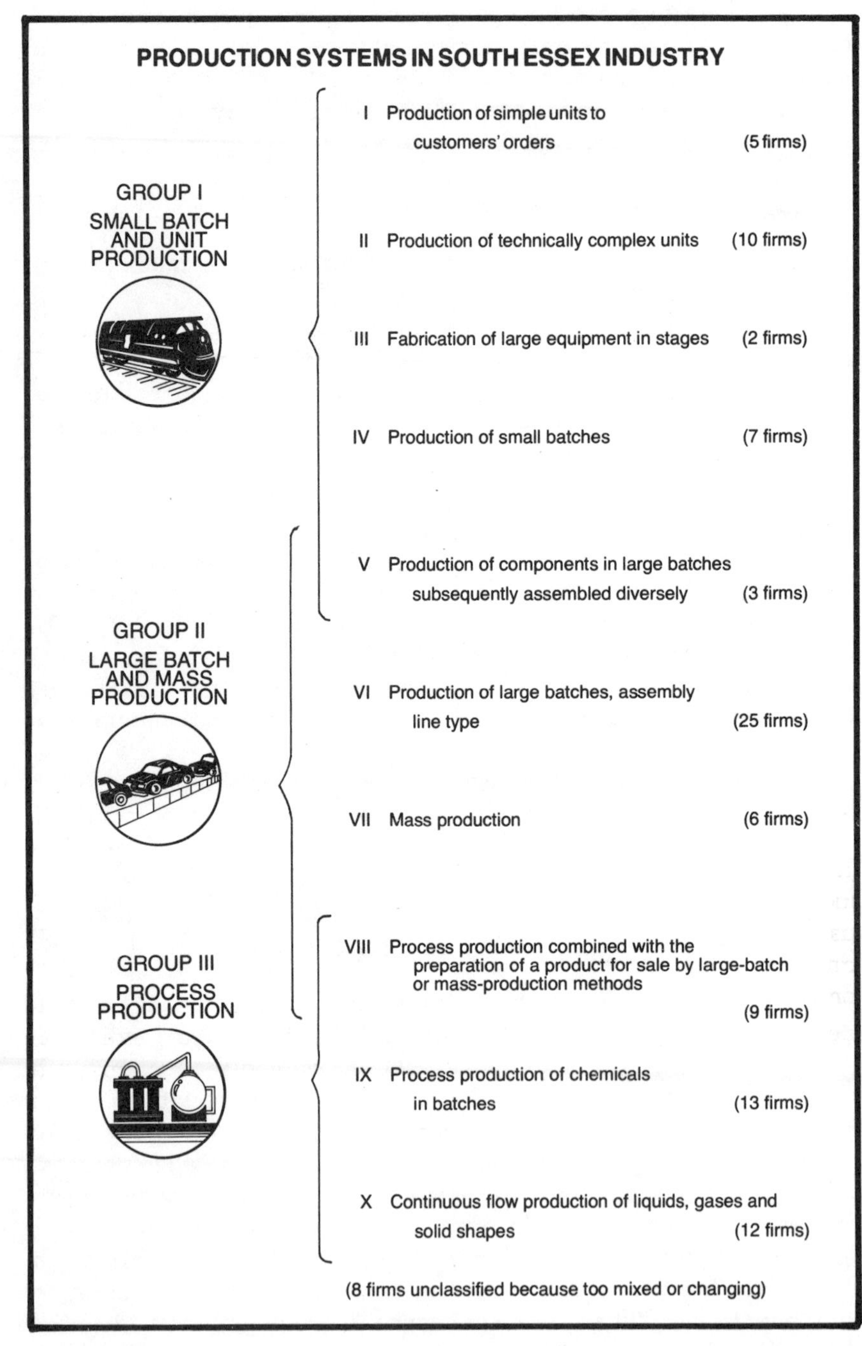
PRODUCTION SYSTEMS IN SOUTH ESSEX INDUSTRY
GROUP I
SMALL BATCH AND UNIT PRODUCTION
GROUP II
LARGE BATCH AND MASS PRODUCTION
GROUP III
PROCESS PRODUCTION
I Production of simple units to customers' orders (5 firms)
II Production of technically complex units (10 firms)
III Fabrication of large equipment in stages (2 firms)
IV Production of small batches (7 firms)
V Production of components in large batches subsequently assembled diversely (3 firms)
VI Production of large batches, assembly line type (25 firms)
VII Mass production (6 firms)
VIII Process production combined with the preparation of a product for sale by large-batch or mass-production methods (9 firms)
IX Process production of chemicals in batches (13 firms)
X Continuous flow production of liquids, gases and solid shapes (12 firms)
(8 firms unclassified because too mixed or changing)

Figure 1.

However, technical developments may from time to time enable a firm to achieve its objectives more effectively through a change in its production system, and a large proportion of manufacturing firms in the future are likely to be process firms. Indeed, although large-batch and mass production is regarded as the typical manufacturing system, less than one third of the firms in south Essex are even now in this production group.

Automatic and other advanced techniques, although more appropriate to some systems than others, are not restricted to any one system. Automatic control can be applied most readily to mass production and continuous-flow process production, but even in unit and small-batch production devices for the control of individual machines can be used.

Sixteen of the firms included in the research had introduced some form of automation. In some of them, for example in those canning food and making mill-board, the production system had changed in consequence; in others it had not. Thus automation can be introduced without a change in the production system, and a change in the production system can be introduced without automation.

ORGANIZATION AND TECHNOLOGY

The analysis of the research described in the previous chapter revealed that firms using similar technical methods had similar organizational structures. It appeared that different technologies imposed different kinds of demands on individuals and organizations, and that these demands had to be met through an appropriate form of organization. There were still a number of differences between firms—related to such factors as history, background, and personalities—but these were not as significant as the differences between one production group and another and their influence seemed to be limited by technical considerations. For example, there were differences between managers in their readiness to delegate authority; but in general they delegated more in process than in mass-production firms.

Organization and Technical Complexity

Organization also appeared to change as technology advanced. Some figures showed a direct and progressive relationship with advancing technology (used in this report to mean 'system of techniques'). Others reached their peak in mass production and then decreased, so that in these respects unit and process production resembled each other more than the intermediate stage. Figures 2 and 3 show these two trends.

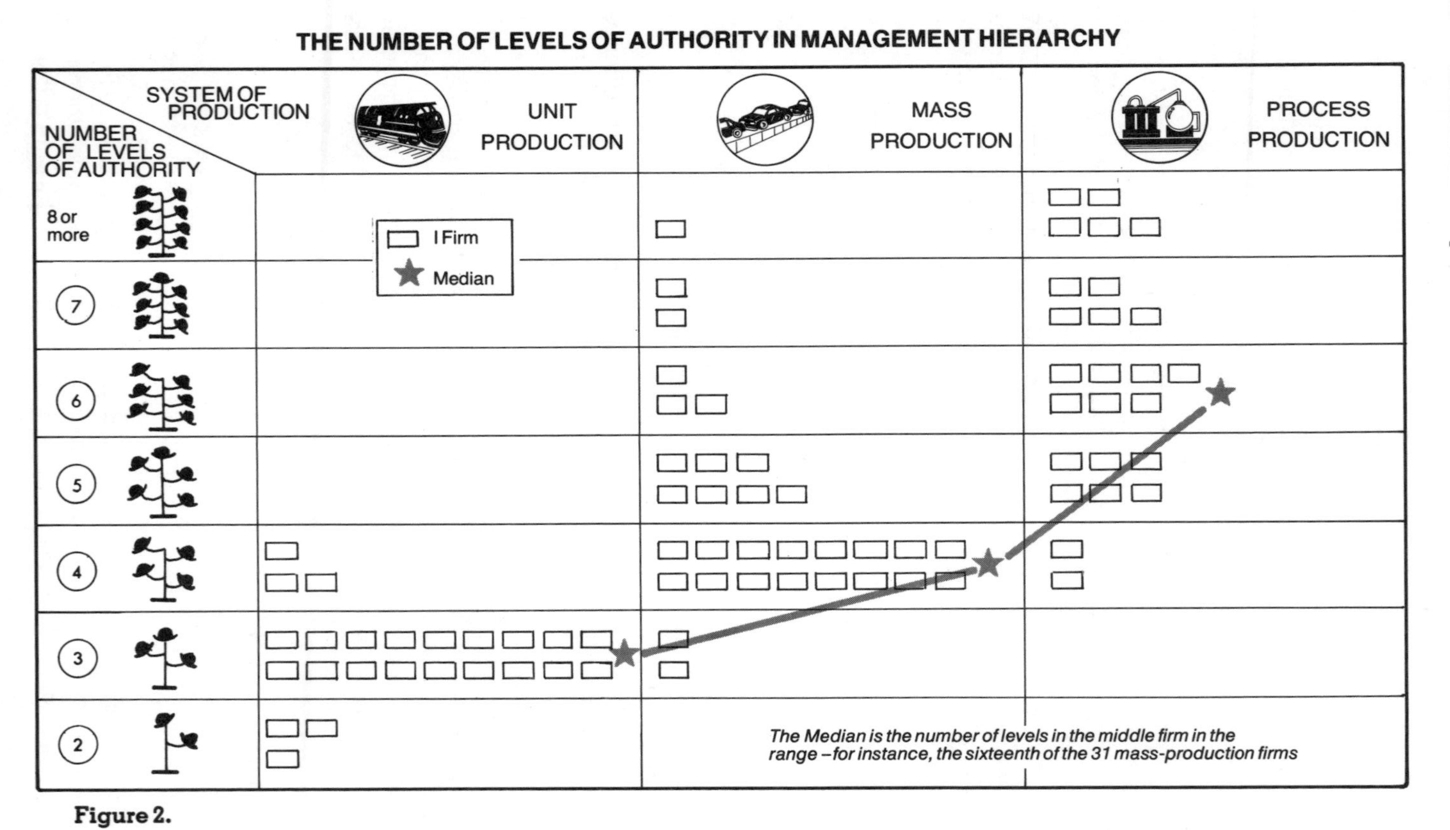
THE NUMBER OF LEVELS OF AUTHORITY IN MANAGEMENT HIERARCHY
SYSTEM OF PRODUCTION
NUMBER OF LEVELS OF AUTHORITY
UNIT PRODUCTION
MASS PRODUCTION
PROCESS PRODUCTION
8 or more
7
6
5
4
3
2
I Firm
Median
The Median is the number of levels in the middle firm in the range – for instance, the sixteenth of the 31 mass-production firms

Figure 2.

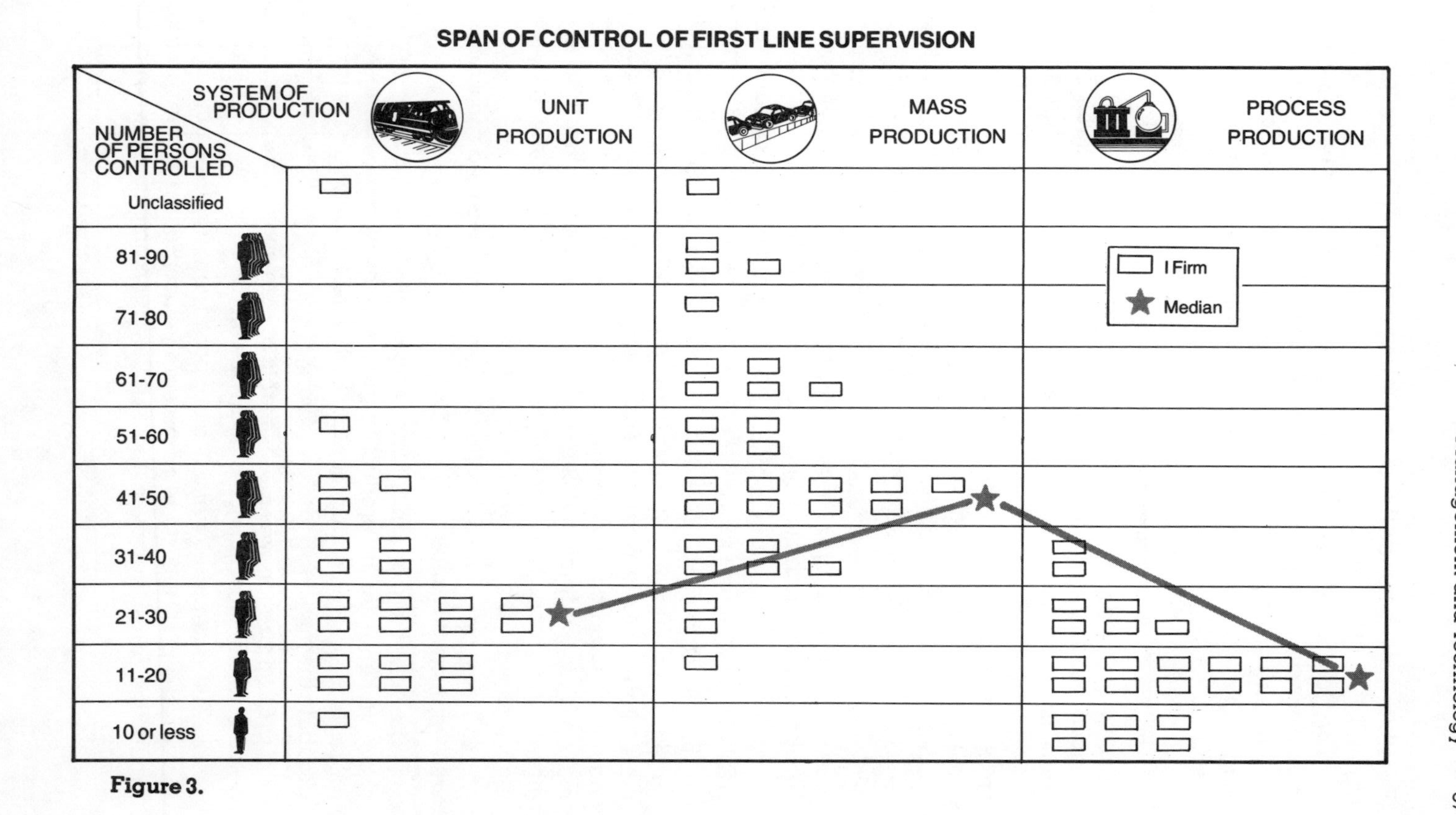
SPAN OF CONTROL OF FIRST LINE SUPERVISION
SYSTEM OF PRODUCTION
NUMBER OF PERSONS CONTROLLED
UNIT PRODUCTION
MASS PRODUCTION
PROCESS PRODUCTION
Unclassified
81-90
71-80
61-70
51-60
41-50
31-40
21-30
11-20
10 or less
I Firm
Median

Figure 3.

(Details are given for the three main groups of production systems. See Figure 1.)

The number of levels of authority in the management hierarchy increased with technical complexity. (See Figure 2.)

The span of control of the first-line supervisor on the other hand reached its peak in mass production and then decreased. (See Figure 3.)

The ratio of managers and supervisory staff to total personnel in the different production systems is shown in some detail in Figure 4 as an indication of likely changes in the demand for managers as process production becomes more widespread. There were over three times as many managers for the same number of personnel in process firms as in unit-production firms. Mass-production firms lay between the two groups, with half as many managers as in process production for the same number of personnel.

The following characteristics followed the pattern shown in Figure 2—a direct and progressive relationship with technical complexity.

1. *Labour costs* decreased as technology advanced. Wages accounted for an average of 36 per cent of total costs in unit production, 34 per cent in mass production and 14 per cent in process production.

2. *The ratios of indirect to direct labour* and of administrative and clerical staff to hourly paid workers increased with technical advance.

3. *The proportion of graduates* among the supervisory staff engaged on production increased too. Unit-production firms employed more professionally qualified staff altogether than other firms, but mainly on research or development activities. In unit-production and mass-production firms it was the complexity of the product that determined the proportion of professionally qualified staff, while in process industry it was the complexity of the process.

4. *The span of control of the chief executive* widened considerably with technical advance.

The following organizational characteristics formed the pattern shown in Figure 3. The production groups at the extremes of the technical scale resembled each other, but both differed considerably from the groups in the middle.

1. *Organization was more flexible* at both ends of the scale, duties and responsibilities being less clearly defined.

2. The amount of *written, as opposed to verbal, communication* increased up to the stage of assembly-line production. In process-production firms, however, most of the communications were again verbal.

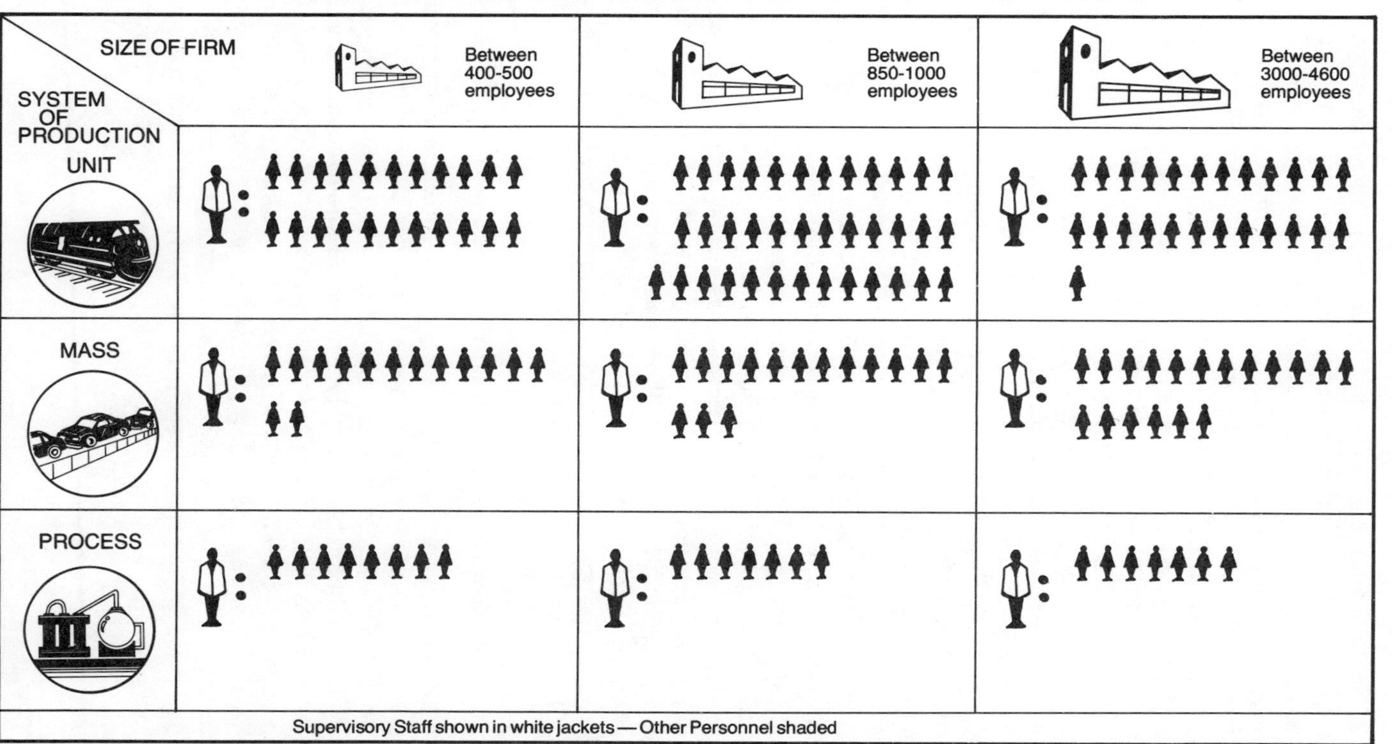
THE RATIO OF MANAGERS AND SUPERVISORY STAFF TO OTHER PERSONNEL
SIZE OF FIRM
SYSTEM OF PRODUCTION
Between 400-500 employees
Between 850-1000 employees
Between 3000-4600 employees
UNIT
MASS
PROCESS
Supervisory Staff shown in white jackets—Other Personnel shaded

Figure 4.

3. *Specialization between the functions of management* was found more frequently in large-batch and mass production than in unit or process production. In most unit-production firms there were few specialists; managers responsible for production were expected to have technical skills, although these were more often based on length of experience and on 'know-how' than on scientific knowledge. When unit production was based on mass-produced components more specialists were employed however. Large-batch and mass-production firms generally conformed to the traditional line-and-staff pattern, the managerial and supervisory group breaking down into two sub-groups with separate, and sometimes conflicting, ideas and objectives. In process-production firms the line-and-staff pattern broke down in practice, though it sometimes existed on paper. Firms tended either to move towards functional organization of the kind advocated by Taylor[5], or to do without specialists and incorporate scientific and technical knowledge in the direct executive hierarchy. As a result, technical competence in line supervision was again important, although now the demand was for scientific knowledge rather than technical 'know-how'.

4. Although production control became increasingly important as technology advanced, *the administration of production*—what Taylor called 'the brainwork of production'—was most widely separated from the actual supervision of production operations in large-batch and mass-production firms, where the newer techniques of production planning and control, methods engineering and work study were most developed. The two functions became increasingly reintegrated beyond this point.

The Effect of Technology upon Human Relations

The attitudes and behaviour of management and supervisory staff and the tone of industrial relations in the firms also seemed to be closely related to their technology. In firms at the extremes of the scale, relationships were on the whole better than in the middle ranges. Pressure on people at all levels of the industrial hierarchy seemed to build up as technology advanced, became heaviest in assembly-line production and then relaxed, so reducing personal conflicts. Some factors—the relaxation of pressure, the smaller working groups, the increasing ratio of supervisors to operators, and the reduced need for labour economy—were conducive to industrial peace in process production. Thus, although some managements handled their labour problems more skilfully than others, these problems were much more difficult for

[5]Op. cit.

firms in the middle ranges than those in unit or process production. The production system seemed more important in determining the quality of human relations than did the numbers employed.

Size and Technology

No significant relationship was revealed between the size of the firm and the system of production. There were small, medium, and large firms in each of the main production groups.

There were firms which employed relatively few people and yet had all the other characteristics of a large company, including a well-defined and developed management structure, considerable financial resources, and a highly paid staff with considerable status in the local industrial community. This was particularly true of the smaller process-production firms. Some of these employed less than 500 people but had more of the characteristics of large-scale industry than unit- or mass-production firms with two or three times as many employees. As indicated already the ratio of management staff to the total number employed was found to increase as technology advanced. It appeared also that the size of the management group was a more reliable measure of the 'bigness' of a firm than its total personnel.

Moreover, although no relationship was found between organization and size in the general classification of firms, some evidence of such a relationship emerged when each of the production groups was considered separately. For example, in the large-batch and mass-production group the number of levels of authority and the span of control of both the chief executive and the first line supervisor both tended to increase with size.

Structure and Success

Again, no relationship between conformity with the 'rules' of management and business success appeared in the preliminary analysis of the research data. The twenty firms graded as outstandingly successful seemed to have little in common.

Table II. Production Systems Analysed by Number Employed

PRODUCTION SYSTEM	NUMBER EMPLOYED: 101-250	251-1000	OVER 1000	TOTAL NUMBER OF FIRMS
Unit	7	13	4	24
Mass	14	12	5	31
Process	12	9	4	25
Totals	33	34	13	80

When, however, firms were grouped on a basis of their production systems, the outstandingly successful ones had at least one feature in common. Many of their organizational characteristics approximated to the median of their production group. For example, in successful unit-production firms the span of control of the first line supervisor ranged from 22 to 28, the median for the group as a whole being 23; in successful mass-production firms it ranged from 45 to 50, the median for the group being 49; and in successful process-production firms it ranged from 11 to 15, the median for the group being 13. (See Figure 3.) Conversely the firms graded as below average in most cases diverged widely from the median.

The research workers also found that when the 31 large-batch- and mass-production firms were examined separately there was a relationship between conformity with the 'rules' of management and business success. The medians approximated to the pattern of organization advocated by writers on management subjects. Within this limited range of production systems, therefore, observance of these 'rules' does appear to increase administrative efficiency. This is quite understandable because management theory is mainly based on the experience of practitioners in the field, much of which has been in large-batch and mass-production firms. Outside these systems, however, it appears that new 'rules' are needed and it should be recognized that an alternative kind of organizational structure might be more appropriate.

CONCLUSIONS

How useful is the foregoing analysis of the demands imposed by the techniques of the production system? What contribution can it make to the study of management organization?

Firstly, the follow-up studies provided explanations of several organizational differences between production groups revealed by the background survey. Secondly, the analytical methods used isolated the forces at work in firms where technical innovation had resulted in disturbances in management organization.

The studies confirmed that variations in organizational requirements between firms are nearly always linked with differences in their techniques of production. For example, differences in two large-batch-production firms of approximately equal size were traced to the fact that one of them, although mainly producing large batches, also made a few articles to customers' individual requirements.

Thus it was possible to trace a 'cause and effect' relationship between a system of production and its associated organizational pattern and, as a result, to predict what the organizational requirements of a firm are likely to be, given its production system. For example, the following features can be traced to the technology of each system of production: a co-ordination of functions and centralization of authority in unit production; an extensive specialization and delegation of authority in mass production; and in process industry a specialization between development, marketing and production, combined with integration within each function and the co-operative character of decision-making.

The background survey showed that the successful firms approximated to the median[6] of the group in which they had been placed. This indicates that the medians for each group represent a pattern of organization appropriate to the technology of that group.

Building Up Organization

Are those responsible for building the organization in successful firms consciously aware of the demands of the technical situation, and does this affect their decisions? The findings suggest it is unlikely. In only three firms was there any definite evidence that organization was determined by a systematic analysis of 'situational demands'.

Organization appeared to grow in response to a number of stimuli. The 'organization conscious' firms tended to draw on the concepts of management theory, irrespective of how appropriate they were to the technical situation. Fashion was another important factor. Materials controllers and industrial engineers were becoming popular at the time of the research, and it was interesting to see how they spread from one firm to another. Moreover, organization had been modified to some extent in every firm to accommodate individuals—'empire builders' who distorted the pattern in their search for status, and misfits for whom some sinecure had to be found. These distortions often continued long after the people concerned had died or left the firm.

But although 'situational demands' did not determine formal organization, they appeared to have considerable influence on spontaneous or informal development.[7] In a number of firms formal organization did not satisfy 'situational demands' adequately, while informal organization did. Social scientists generally believe that a wide divergence between formal and informal organization is undesirable because it creates tension and conflict. Some of the firms in which organizational

---

[6]See Figure 2.

[7]Formal organization implies the stable and explicit patterns of prescribed relationships in the firm, while informal organization implies the pattern that actually emerges from day-to-day operation.

structure appeared to diverge from 'situational demands' might therefore be less successful because the informal organization deviated from the formal rather than because the demands of the technical situation were unsatisfied.

The most successful firms are thus likely to be the 'organization conscious' firms, in which formal organization is appropriate to the technical situation. Next would come the less 'organization conscious' firms, where informal organization mainly determines the pattern of relationships. The least successful firms are likely to be the 'organization conscious' firms, where formal organization is inappropriate and deviates from informal organization. The research workers found instances of lack of success in 'organization conscious' process-production firms, where formal organization had been developed on a basis of traditional principles of administration which were more appropriate to large-batch and mass production.

'Situational Demands' and Technical Change

The systematic analysis of 'situational demands' might well be used to predict the effects of technical changes on management structure and to plan organizational and technical change simultaneously.

The research showed that some technical changes have more effect on organization than others; their nature can be analysed systematically.

Examples were found of several different kinds of technical change. Changes from unit to batch and mass production were mostly associated with changes in objectives; a more standardized product was to be manufactured. Where, however, individual units were being produced on a basis of mass-produced, standardized parts, the production systems had changed while the objectives remained as before. The introduction of continuous-flow automatically controlled processes into the manufacture of solid shapes—for example, paper-board and mill-board, and food canning and packaging—had changed systems of production but not their objectives.

Two examples of transfer lines were found. In one firm this machinery had been introduced to produce barley-sugar sticks and had resulted in a change from batch to process production. But in the motor-car industry, transfer machines, although on a much larger scale, had not fundamentally changed the system of production; they produced only a very small percentage of the components for assembly.

To sum up: technical changes not associated with changes either in objectives or the production system would be unlikely to create very much disturbance in the organizational pattern. Where, however, the proposed technical change appeared to be likely to create new 'situational demands', these could be foreseen by a systematic analysis of the new technology. Technical developments of the last twenty years, startling as they are in many ways, have not led to any entirely new system of production. They need not give rise, therefore, to organizational problems for which at least partial solutions cannot be found in the accumulated experience of industry.

Implications for Teaching

It was hoped that this research would produce findings on which the management course offered in the South East Essex Technical College could be appraised. At first sight this report may suggest that these courses have limited usefulness and can in some circumstances be misleading to students. The danger lies in the tendency to teach the principles of administration as though they were scientific laws, when they are really little more than administrative expedients found to work well in certain circumstances but never tested in any systematic way. This does not mean, however, that management theory has no value; it contains important and valuable information and ideas, provided its limitations are recognized and its principles subjected to critical analysis.

Management studies can so far identify symptoms and remedies. Alleviation of symptoms is useful in itself, but it is only through diagnosis that a physician can either be sure he is prescribing the right treatment or make any useful contribution to existing knowledge of disease. Thus it is important that in alleviating symptoms physicians do not neglect the problems of diagnosis. In the field of management studies many more descriptive accounts like that given here, of the circumstances in which different administrative expedients have proved successful, are required to supplement traditional teaching.

Technology, Territory, and Time: The Internal Differentiation of Complex Production Systems[1]

Eric J. Miller

The present paper explores the principles of differentiation of operating units within a complex system. Part I describes these principles or dimensions of differentiation in the context of transition from a primary or simple production system to a complex system. An analysis is made of the forces that may act on a simple system to transform it into a complex system containing differentiated operating and managing systems. There is discussion of what happens inside the system when such a transition takes place. In Part II the dimensions are examined in relation to the order of differentiation in a multi-level, multi-shift complex system. Finally, Part III deals with the patterns of relationships—and therefore the problems of management—which are inherent in the structure of certain typical production systems, according to the basis of internal differentiation into subsystems and the nature and degree of sub-system interdependence.

In Lewinian terms, we are dealing here with the 'foreign hull' of the life space of the individual—'facts which are not subject to psychological laws but which influence the state of the life space' (8, p. 216; see also pp. 68-74).

---

[1]I am indebted to my colleagues in the Tavistock Institute of Human Relations for many helpful comments on and criticisms of earlier drafts of this paper.

Extracted from Eric J. Miller, "Technology, Territory, and Time: The Internal Differentiation of Complex Production Systems." HUMAN RELATIONS, 1959, *12*, 245-272. Used with permission of Plenum Publishing Corporation.

PART I: THE TRANSITION FROM SIMPLE TO COMPLEX SYSTEM

The typical simple production system in industry is the primary work-group. Elsewhere it appears in the small workshop, the retail shop, the service station, and so forth. The essential feature of such a system is that management is inherent in relationships within the group: either there is no recognized leader at all (as is the case in some mining groups (5, 14), or, if there is one, he spends all or most of his time working alongside the other members of the group on tasks comparable to theirs. His contribution to the output of the group tends to be directly productive rather than indirect and facilitative.

Herbst has described certain characteristics of simple and complex behaviour systems (3). He finds that one significant criterion of a simple system is that the relationship between input, size, and output is linear. (Input and output are here measured in money rather than goods.) In small retail shops, for example, the total amount paid in wages (i.e. input) increases at a linear rate with sales turnover (output) achieved, while sales turnover increases at a linear rate with the size of the shop, measured by the number of persons employed. In a complex system the relationship is non-linear: 'The presence of an administrative unit concerned with ongoing activities increases the rate at which sales turnover increases with size of the organization, and . . . the loss incurred by withdrawing personnel from production tasks decreases as the organization becomes larger' (3, p. 344). Herbst has this to say about the transition from a simple to a complex system:

> 'As the size of the simple system increases, and depending also on the extent of both its internal and external linkages, more and more work has to be carried out on the co-ordination of component functioning, so that a critical boundary value with respect to size is reached, beyond which intrinsic regulation breaks down. An increase in size beyond this point will become possible by differentiating out a separate integrating unit, which takes over the function of both control and co-ordination of component units, thus leading to a transition from a simple to a complex system. The point at which intrinsic regulation breaks down will be determined by the effectiveness of the organizational structure. The less efficient the organizational structure happens to be the earlier the point at which intrinsic regulation breaks down' (3, pp. 337-8).

In other words, three critical factors in the transition are size, complexity, and efficiency. Herbst here seems to be implying that the efficiency of a simple system is measured by the extent to which the system can tolerate increased size and complexity without throwing up a differentiated management function. However, if efficiency is measured by the ratio of output to size—a ratio that Herbst himself uses elsewhere in the same article—then it would appear that this assumption might not always be justifiable. If, for example, a simple system of given size could secure a greater output by becoming reorganized as a complex system of the same size then its persistence as a simple system would be relatively inefficient. 'Resilience' might therefore be a more appropriate term than 'efficiency' to describe the capacity of a simple system to withstand pressures—both external and internal—towards transformation into a complex system. This would be an omnibus term embracing a number of factors of small group functioning which counter the effects of increased size and complexity. Some of these factors are considered later in this section.

Size by itself is not a critical factor. Apart from the pair, groups of six to twelve are often said to be the most stable, in both psychotherapeutic and other situations (10, pp. 36-7). Fissiparous forces tend to develop in groups outside this optimum range, but there is no known maximum number beyond which emergence of a full-time management function is inevitable. Much depends on the need for differentiation that is intrinsic in the task of the group and in the way the task has to be, or is being, performed. Herbst notes, for example, that in independent retail shops a differentiated administrative function tends to appear when the staff numbers around five, whereas shops that belong to a retail chain retain the characteristics of simple systems until the size reaches about nine. Certain services are supplied to the latter by the larger organization to which they belong. In the Durham coalfield autonomous groups of forty-one have been shown to be effective (14). They have internally structured controls and services and lack any overtly recognized and titled leader. These groups are further discussed below.

Complexity may be considered in terms of Rice's import-conversion-export formulation. Imports into the system and exports from it may become more diverse. Complexity of the conversion process is likely to increase through diversification of input or output or both, or through a change in the techniques or rates of production.

Before considering these factors of size and complexity in more detail, it seems necessary to stress that *an essential preliminary to differentiation of a managing system is the formation of sub-systems with discrete sub-tasks within the simple system*. Role-relationships cluster around the sub-tasks; such clusters of relationships become potential sub-systems; and areas of less intensive relationships become potential boundaries between sub-systems. Clustering may be functional for sub-task performance, but the associated discontinuities between clusters may be dysfunctional for integrated performance of the total task. It becomes a function of a differentiated managing system to compensate for these discontinuities. Management mediates relationships among the lower-order systems, which constitute the higher-order system, in such a way as to ensure that the sub-tasks performed by the sub-units add up to the total task of the whole unit.

If the principles of differentiation of sub-systems can be identified, then the effects of changes of size and complexity can be more clearly understood; and furthermore the notion of 'resilience'—the capacity to withstand, without sacrifice of efficiency, the pressures towards creation of a differentiated managing system—will become less vague.

It is postulated here that there are three possible bases for clustering of role-relationships and thus for the internal differentiation of a production system. These are technology, territory, and time. Whenever forces towards differentiation operate upon a simple production system, it is one or more of these dimensions that will form the boundaries of the emergent sub-systems and will provide the basis for the internal solidarity of the groups associated with them.

'Technology' here is given a broad meaning. It refers to the material means, techniques, and skills required for performance of a given task. Differentiation of the import, conversion, and export systems (the purchasing, manufacturing, and selling of an industrial unit) is in this sense a technological differentiation; so also is differentiation of phases of the conversion operation (successive manufacturing processes), or specialization in buying or selling particular commodities. The greater the diversity of technologies used within a group, the stronger the forces towards differentiation of fully-fledged sub-systems—especially when the skills of some members are so specialized that others cannot aspire to have them or even comprehend them, and interchange of roles between members of the total group becomes impracticable.[2] Increase in technological complexity or diversity tends to have this effect even though the quantum of input and output remains unchanged. It may even occur where the size of the system, in terms of the number of roles, is reduced.[3]

The dimension of territory is straightforward: it relates to the geography of task performance. An increase in the staff of a retail shop from three persons to five may not precipitate formation of a differentiated

---

[2]The obverse point was made by Rice (10, pp. 37-9), who postulated that small work-groups in modern mechanized industry usually require sufficient variety of roles (implying some technological differentiation) as to need some internal structuring, but should not have so much specialization as would lead to formation of inflexible sub-groups.

[3]In her study of industrial firms in south Essex, Joan Woodward noted that 'the number of levels of authority in the management hierarchy increased with technical complexity', while 'the span of control of the first-line supervisor . . . decreased' (15, p. 16).

management function. If, however, the two extra persons are employed to start a branch store—if, in other words, two potential sub-systems are formed, spatially separated from one another—then the forces towards differentiation will be greatly increased. Physical separation is not essential to produce this result, but a sharp physical boundary of some kind is probably necessary before territory by itself can become a basis of sub-system differentiation within a simple production system. Identification of the group with its territory is of course a basic feature of all human societies and is found too among many of the higher mammals. Even boundaries that are imperceptible to an external observer may have highly charged emotional significance for the members of the groups they divide—especially when territorial differentiation is reinforced by technological differentiation. Technology, indeed, seems to seek the support of territory, and only seldom stands by itself as a differentiating factor. (In most parts of India, castes which are differentiated from one another by their traditional occupations are also segregated spatially, living in different parts of the village or in different villages.)

The third dimension—time—is more commonly relevant in increasing the levels of differentiation in an already differentiated complex system, but may also reinforce an increase in size in bringing about the transition from a simple system to a complex one. Forces towards differentiation probably begin to develop when the requirements of task performance are such that the length of the working day or working week of the group exceeds the working period of any individual member. This factor of time is of course most pronounced in multi-shift systems. As in the case of territorial separation, sub-systems tend to emerge with well-defined boundaries, which in this case are based on time separation.

The sub-systems and associated groupings described in the preceding paragraphs are those that are intrinsic to the structure of the task. Task structure is assumed to be inseparable from the type of technology and specialization involved, from the geography of the territory in which the task is performed, and from the time-scale of task performance—though within these limiting factors alternative structures may be possible.

Among the persons manning the roles of a production system other groupings may occur, based on propinquity, sex, age, religion, race, and many other principles of association, and on occasion these groupings and related cleavages, perhaps by their coincidence with task-oriented groupings, may accelerate differentiation; or, if they cut across these groupings, they may retard it. It is the task-oriented sub-systems themselves, however, which are relevant to task performance. These seem invariably to be differentiated by technology, territory, time, or some combination of these.[4] Production systems can probably not be satisfactorily broken down into sub-systems on any other basis.[5]

If territory, technology, and time, singly or in combination, provide the basis for differentiation into task-relevant sub-systems, the capacity of a simple system to tolerate growth and remain efficient without becoming transformed into a complex system is apparently related to two main factors: (a) mobility or fluidity, and (b) sub-system interdependence.

If individual members move frequently from one sub-system to another, so that there are no permanent sub-groups of workers coinciding with the task sub-systems, then the simple system will have greater capacity to tolerate an increase in size or complexity. Such movement compensates for discontinuities between sub-systems. Secondly, the more immediately and directly performance of the task of each sub-system depends upon the performance of all the other sub-systems, the more likely is the total simple system to remain viable in the face of forces towards differentiation. (Without some task interdependence it is, of course, not a production system but an assembly or aggregate of individuals.)

---

[4]Since drafting this paper I have seen Gulick's five-fold classification of the ways in which work in an organization can be grouped: by purpose (cf. by product), by process, by clientele, by place, and by time (2). My 'technological' category would embrace the first three of these. There seems, however, to be a conceptual distinction between the grouping of work and the differentiation of socio-technical systems, and in several respects the similarity between the 'scientific-management' theories of Gulick, Urwick, *et al.* and the aspect of socio-technical-system theory elaborated in this paper is only superficial. One difference is that (as, for example, March and Simon (9) have pointed out) Gulick's models do not take human motivations into account: he is dealing with management of technical systems, not of *socio*-technical systems. In Gulick's scheme, principles of association extraneous to work are obstructions to be overcome in establishing an essentially static and inflexible organization; in the more dynamic socio-technical-system theory they are a relevant part of the total reality situation. In general, the Gulick type of organizational theory does not lend itself to predictions about behaviour and relationships.

[5]The women's services in the forces may at first sight appear to be an exception to this rule. Closer examination shows, however, that these do not constitute production systems within which the members are interrelated and integrated by performance of a specific task. The production systems in the forces are the training and fighting units to which members of the women's services are attached (in much the same way as medical and signals personnel are attached to an infantry battalion) and to the task of which they contribute. In terms of the tasks of these units, the women may do jobs which are similar or dissimilar to jobs of men in the same unit; but in either event it is the technological specialization which is relevant to the total task, and their sex is incidental.

Exceptionally large simple production systems occur in longwall coal-mining, in a form of composite working described by Trist and Murray (14) and Higgin (5). As mentioned earlier, some of the composite groups have forty-one members, working over three shifts. Both 'resilience' factors operate strongly in these groups, which are internally differentiated by both technology and time. Although the individual sub-systems have well-defined tasks, mobility between the sub-systems allows many or all of the members to view inter-sub-system relationships from the perspective of the total system rather than from the perspective of the sub-systems they happen to belong to at any one time. Close reciprocal interdependence, necessary in these mining groups for achieving the total task, evidently helps to reinforce this global perspective.

It may well be that it is not the number of persons that limits the maximum size of a simple production system, but the number of sub-systems. (A sub-system may consist of either an individual or a sub-group.) Certainly, complexity in task structuring can actually contribute to the cohesion of large simple systems. Where there is a number of sub-systems interdependent in more than one direction, the complex conditions of equilibrium can be a substitute for a differentiated management function. It is the very lack of such complexity built in to the task that helps to lower the threshold of 'resilience' in less structured simple production systems. Internal structuring for which the primary task does not cater is sought in other groupings (based on age, sex, etc.), implying involvement in other tasks that to a greater or lesser extent conflict with the primary task for which the system was constituted. In some cases it may be possible to use these factors of resilience and to restructure roles in such a way as to postpone the emergence of a differentiated managing system.

It can be inferred that, in any expanding or changing system in which no such restructuring has occurred, there is an optimum or 'natural' stage for creating a new level of management. This is applicable equally to the initial transition from a simple system to a complex system and to the addition of a new level to an already complex hierarchical system.

Premature differentiation is uneconomic because the cost of adding a specialized administrative function is greater than the gain from any increase in efficiency that results. As sub-systems have not yet been crystallized by task differentiation, government is more efficiently contained as an undifferentiated internal function. Indeed, extrinsic government, if imposed prematurely, may tend to be more destructive than integrative. (This is the kind of situation in which the internal collaborative relationships, which before the change have been used constructively for task performance, are likely to be mobilized destructively against the imposed external management: cf. Trist and Bamforth (13).)

Postponement of differentiation of the management function beyond the optimum stage also leads to a decline in the efficiency of the system, but for a different reason. The energies of group members, instead of being devoted to the primary task, are increasingly diverted to the task of holding the group together in the face of the fissiparous forces of sub-group formation and of differentiation. This is especially likely to happen if there is imbalance in the pattern of sub-system interdependence. Individuals experience conflict between identification with an emergent sub-group and identification with the total group. Only the creation of a new level of management which allows the sub-systems to become fully explicit simple systems and which reintegrates them as parts of a higher-order system, permits the energies of the members to revert to primary task performance.

Herbst uses the input-size-output relationship as an index for measuring the level of behaviour systems and for diagnosing whether a given system is simple or complex (3). The reverse approach may also be useful. If a production system, which is known to have the structural characteristics of a simple system, increases in size, and if this expansion is unaccompanied by a linear increase in output, then (other things being equal) it is worth investigating whether the system has passed the optimum stage for differentiation—either because the sub-systems are in a stage of disequilibrium or because of emergence of sub-groups unrelated to the primary task of the system. The same possibility may exist if a simple system, remaining constant in size, shows over a period of time a declining output. Equally, if a structural transition from a simple to a complex system is not accompanied by the kind of change in size-output ratio predictable from a Herbst-type formula for systems of that kind, then it is possible that differentiation of the managing system has been premature.

PART II: STRUCTURE OF COMPLEX PRODUCTION SYSTEMS

The forces towards transforming a simple system into a complex system, or towards increasing the levels of differentiation in a system that is already complex, are not only of theoretical interest to social scientists, but also of practical interest to those concerned with management. It has already been suggested, for instance, that working efficiency and cost are likely to be adversely affected if the timing of a change in response to the accumulating forces towards differentiation is not opportune. A second cause of inefficiency may lie in an inappropriate choice of the basis of differentiation into sub-units.

In the initial transition from a simple to a complex system, the basis of differentiation is usually directly traceable to the forces leading to differentiation. Consider the example of a small privately owned workshop that manufactures simple components, all of the same kind, for the automobile industry. Raw materials are delivered and the finished products removed by the company it supplies. Administration takes up little of the owner's time and he himself works at a bench alongside his employees. This is the typical simple production system. Let us imagine that demand grows, and because of lack of space for expansion the owner acquires two more workshops in the vicinity. If the three workshops are sufficiently far apart, the owner is likely to spend less of his time at the bench and to take a nearly full-time managerial role. The three workshops then become three simple operating systems within a complex production system. In other words, territorial expansion has led to differentiation and it is territorially demarcated sub-units that are explicitly recognized.

Alternatively, the expansion might have been achieved by adding two more shifts in the original workshop. The shifts would then become the recognized sub-units, and, because of the need for control and co-ordination over the twenty-four hours, the owner would again take a full-time managerial role.

We now have to consider what happens when additional forces towards differentiation operate on a production system that has already become complex and there is the prospect of extending the hierarchy by further differentiation. Here again the forces themselves will dictate the new basis for differentiation, but not necessarily the level at which it will occur.

Reverting to our example, let us now suppose that further expansion requires all three workshops to run on three shifts. Each shift in each workshop is now likely to develop into a simple sub-system, and sooner or later the owner-manager will be compelled to realize that there are nine workshop-shifts to be managed, instead of merely three workshops on one shift, or three shifts in one workshop (Figure 1).

Time Differentiation

Workshop A Shift I	Workshop A Shift II	Workshop A Shift III
Workshop B Shift I	Workshop B Shift II	Workshop B Shift III
Workshop C Shift I	Workshop C Shift II	Workshop C Shift III

Territorial Differentiation

Figure 1. Sub-Units Differentiated by Both Territory and Time

Increase in the number of sub-units does not, of course, necessarily lead to further differentiation and to an increase in the number of levels. If, for example, output had been tripled by expanding from three workshops to nine, instead of by adding more shifts, the additional simple production systems so created could have become explicit without necessarily over-extending the span of the overall manager's command. Even in the present example, it might be practicable to maintain direct control of the nine sub-units, perhaps by employing additional staff for time-keeping and recording production—that is, by increasing the size of the managing system without adding to the number of levels in the hierarchy. However, since the sub-systems in this case are differentiated and interdependent along two dimensions (territory and time) that cut across each other, and therefore have to be co-ordinated along these two dimensions, it is likely that an additional level of management will be interposed.

The owner is now faced with a choice. He may introduce the new level by managing the three territories (workshops) through three foremen, delegating to the foreman in each workshop the task of co-ordinating the three shifts within it. Alternatively, he may elect to undertake co-ordination of the three shifts himself by appointing three shift foremen, each of whom is responsible for the work on one shift in all three workshops.[6] The fact that territorial differentiation preceded the addition of shifts by no means presupposes that, in the management hierarchy, territorial differentiation need occur at a higher level than differentiation by time.

It is now necessary to consider this choice in more detail. In fact, it is a real choice only in so far as territory and time are equally salient in differentiating the nine simple systems from one another. In terms of task relationships, this is so only when one shift in one workshop is equally interdependent with other shifts in the same workshop and with the corresponding shift in other workshops. Workshop *A* Shift I (*A* I) belongs then to two larger systems: it is part of the '*A*' system, within which the other systems are *A* II and *A* III, and it is part of the 'I' system, within which the other systems are *B* I and *C* I (cf. Figure 1). In the situation of equal interdependence,

$$R(A\,I, A\,II, A\,III) = R(A\,I, B\,I, C\,I),$$

where R is a measure of task interrelatedness between the simple systems. Such an equilibrium may make it possible for the nine workshop shifts to be managed directly without interposing a new level of differentiation.

---

[6]It was A. K. Rice who first drew my attention to this kind of choice, to which he also refers in his recent book (10, pp. 177 and 200-1).

We have seen that the formation of sub-systems with discrete sub-tasks is a necessary preliminary to transition from a simple to a complex system. Similarly, in an expanding complex system, the clustering of sub-systems precipitates an additional level of differentiation, in which the clusters are acknowledged as explicit systems of a higher order than the constituent sub-systems.

When two dimensions of differentiation are involved, with two implicit sets of systems cutting across one another, it is seldom that they actually have equal salience. Task relationships generally draw the basic units into the orbit of one system more strongly than into the other, and so dictate the lines of higher-order and lower-order differentiation. Furthermore, even if task-oriented interrelations themselves do have equal salience, other factors may tend to tilt the balance one way or the other. Persons who share a compact territory over three shifts, for example, may feel more strongly identified than those who share the same shift-timing over dispersed territories. Alternatively, if the dispersal is limited, going to work at the same time, and hence sharing free time, may lead to closer identification.

Failure to differentiate on the appropriate basis will create stress in relationships, because the natural groupings inherent in the structure of task performance will run counter to the groupings dictated by the formal organization. Formal boundaries will cut through these natural groupings. This will inhibit development of solidarity in the formal units, with consequent lowering of work satisfaction and morale. In general, we can suggest that to the extent that the formal structuring deviates from the reality of the task situation, whether in the basis for differentiation or in the boundaries of the formal sub-units, to that extent will the management function itself have to multiply and become 'top-heavy' in order to deal with the resultant dysphoria. Additional controls will have to be imposed. This tendency will increase in proportion to the interdependence of the formal units. If on the other hand a unit is appropriately sub-divided in relation to total task performance—if it is cut, so to speak, with the grain and not against it—both the internal management of the constituent sub-units and the overall integration of the total task are likely to require less effort.

Flexibility is not entirely lacking. Imposition of a managing system itself helps to crystallize the selected basis and boundaries of differentiation of operating systems. Therefore, provided that the salience of two dimensions is not too unequal, differentiation at the higher level along the dimension of lower salience may increase the salience of that dimension to a point where it exceeds that of the other. This would not appreciably increase the difficulties of management. Similarly, if prior clustering of sub-units is not too strong, the emergent boundaries can be supplanted by formal boundaries that do not necessarily coincide with them. Such flexibility, however, occurs only in marginal cases.

So far, instances of only two orders of differentiation have been discussed—by territory and time. We have seen that there is, subject to certain limiting factors, a choice between:

(a) first-order differentiation by territory and second-order differentiation by time; and
(b) first-order differentiation by time and second-order differentiation by territory.

A third possibility, provided the salience of the two dimensions is roughly equal, is to accept only one order of differentiation, operating systems being differentiated simultaneously by the time dimension in one direction and by the territorial dimension in the other. This was illustrated in Figure 1, which shows three time sub-divisions and three territorial sub-divisions, making nine sub-units in all. Theoretically there is yet another way of compressing differentiation by two dimensions into one level. This occurs when the two dimensions, instead of operating at right angles, coincide and reinforce one another. Time and territory coincide in this way when shift working is used in highly mechanized road construction. A piece of mobile equipment—the common technology—is operated by one team in one stretch of road on Shift I, by a second team on a fresh territory on Shift II, and so on. In longwall coal-getting, time and technology coincide as differentiating dimensions, territory being undifferentiated: a different technology is used on each of three different shifts on one coal-face. Both these combinations are fairly rare in industry, where it is territory and technology that most frequently coincide as reinforcing dimensions: in manufacturing operations, more often than not, each of a group of technologically differentiated sub-units has its own discrete territory of task performance as well.

When all three dimensions of differentiation occur (if, in the example of the workshop, several products are manufactured in each of the three workshops on three shifts), the theoretical choice of order of differentiation is greatly increased. Assuming that differentiation occurs only once along each dimension, there are six combinations of three levels of differentiation, six more of two, and one of one level. It should be noted that in the seven combinations involving differentiation by more than one dimension at one level, the simultaneous differentiation may be either (a) cross-cutting or (b) coincident and reinforcing.

Where there are more than three levels, at least one dimension will become the basis of differentiation at more than one level. In a large manufacturing concern, for example, there may be first-order differentiation into purchasing, manufacturing, and sales (technology); second-order differentiation of manufacturing into product units (technology, probably reinforced by territory); third-order differentiation of the product units into departments responsible for various phases of the process (again technology plus territory); and so forth. Time differentiation will occur in a multi-shift concern, but a twenty-four-hour command is narrowed down into eight-hour commands at only one level in any segment of a hierarchy. It may nevertheless occur at different levels in different segments of the same total hierarchy.

Very often the internal structure of a large organization is the cumulative result of many small local changes. Adherence to a particular pattern of differentiation adopted in response to one change may limit the possible responses to subsequent changes. In so far as the enterprise is a system, a change in one area will affect other areas. Accordingly, any organizational change must be planned in the context of the total structure, to ensure that it provides for the most efficient performance of the primary task of the enterprise.

To sum up, therefore, any production system, complex or simple, can be defined along the dimensions of territory, technology, and time. A large system is broken down into progressively smaller systems along one or more of these dimensions at each level. The smallest systems are sometimes co-extensive along one or even two dimensions with the overall system, but more often in a manufacturing organization they are shorter along all three dimensions. Each component system, however, has boundaries that serve to separate it from parallel systems, and also boundaries that form part of the higher-order system's boundaries. Work-oriented relations crossing the former boundaries should be more intensive than those that cross the latter; if not, it can be inferred that an inappropriate basis of differentiation has been adopted and that the efficiency of the total system is less than optimal.

PART III: INTERNAL DIFFERENTIATION AND PROBLEMS OF MANAGEMENT

Where a complex production system is differentiated into sub-systems, the total task is also broken down into sub-tasks associated with these sub-systems. As Rice has pointed out, such a hierarchy of tasks may often lead to situations where 'decisions taken within one component system which are consistent with its primary task may appear irrelevant or even harmful in a system of a different order' (10, p. 228). Differentiation into sub-systems therefore throws up a managing system which has the reintegrative function of seeing that the constituent tasks of the sub-systems are so performed that they add up to the overall task of the system as a whole.

It is suggested here that the way in which a task is broken down—in terms of the dimensions along which the sub-systems are differentiated and in terms of the intrinsic interdependence between them—is a major determinant of the kind and quality of management required, including the kinds of control mechanism that will be appropriate. Fundamentally, of course, the dimension along which the system is differentiated at a given level is the dimension along which the major controls have to be exercised to secure reintegration.

Differentiation by territory, technology, and time, singly and in combination, can at any one level take seven different forms—three one-dimensional, three two-dimensional, and one three-dimensional. These are set out in Figure 2. Multi-dimensional differentiation can be reinforcing, cross-cutting, or mixed, though the examples given in Figure 2 are all of reinforcing differentiation: that is, at the level of differentiation in question, each component system is differentiated from every other along both the named dimensions. Examples of cross-cutting and mixed differentiation could also be added.

Differentiated Dimensions	*Undifferentiated Dimensions*	*Examples*
1. Territory	Technology and Time	(a) Separate sections within a factory, or separate factories, making same product (b) Marketing organization; chain of retail stores
2. Technology	Time and Territory	Shipbuilding
3. Time	Territory and Technology	Typical multi-shift structure in process and other industries
4. Territory and Technology	Time	(a) Quasi-independent product units (b) Consecutive manufacturing operations
5. Technology and Time	Territory	Longwall coal-mining
6. Time and Territory	Technology	Mechanized road-making with shift working
7. Territory, Technology, and Time		Milk: collection; processing and bottling; and delivery

Figure 2. The Seven Basic Forms of Differentiation at One Organizational Level. The Examples of Two- and Three-Dimensional Differentiation Given Are of 'Reinforcing' Type. Brief Notes on Three Examples Are Given in the Text.

Types of task dependence have been classified in some detail by Herbst (4) and Emery (1). For present purposes it is relevant to consider the extent to which, at a given level of differentiation, the component systems of a larger system are *co-dependent* on supplies, equipment, and services, and *interdependent* for the attainment of the end-result or goal of the larger system. One or both of these types of dependence may be present. Emery points out that interdependence may be simultaneous or successional, and that successional dependence may be further classified as cyclic, convergent, or divergent. Distinctions can also be drawn between simple and complex dependence and between reciprocal and non-reciprocal.

If the differentiation variables were separately considered in relation to all the dependency variables, the resultant number of combinations would be enormous. Here it will be sufficient to examine the three basic differentiation variables in a little more detail and to discuss a few models that occur fairly frequently in industry. From these the implications of other models can be inferred.

There is one other respect in which the present discussion is restricted. While the basis on which sub-systems are differentiated and the nature of their dependencies are the internal system elements that create a particular pattern of demands on management, it is also a function of management to mediate in certain ways between the system and its environment (which may include successively larger systems of which it is a part), and environmental factors will inevitably impose certain other demands. Such factors, for example, may call for additional control mechanisms within the system. The more complex and diverse these environmental factors are, the greater the number and variety of control and service functions likely to be differentiated within the managing system, and the greater the consequent complexity of intra-system relationships. Here, however, environmental factors are held constant and attention is focused on internal factors relevant to the relationships of a manager with his immediate subordinate group.

Differentiation by Territory

It is characteristic of operating systems differentiated from one another only along the territorial dimension that the output of the total system to which they belong is the added sum of the outputs of the constituent systems. Output from one system can be high, low, or even absent without directly affecting output from the others. In other words, where differentiation is only territorial, interdependence is minimal.

The extent to which the systems are co-dependent—on a single source of supplies, for example, or on centralized service functions—can vary considerably. Spatial segregation can be an important factor here, though not necessarily a determining one. To take the examples given in Figure 2, if the territorially differentiated units are neighbouring sections in the same factory—for example, the series of groups of workers on groups of looms in the textile mills described by Rice (10)—they are likely to draw their input from the same source and to be jointly dependent on a number of centralized services. If, however, the units are separate factories making identical products in different parts of the country, their co-dependence may well be less. Canneries and other food-processing plants are often dispersed in this way in order to be close to agricultural sources of supply. Decentralized control over input is practicable in such cases but is less appropriate where the factories (perhaps dispersed to be close to their markets) share a common and limited source of supply. Co-dependence may also extend to output: the smaller the fluctuation of output permissible in the total system, the greater the centralized control over the outputs of the constituent systems.

Putting it in another way, we can say that where a unit is differentiated into territorial sub-units, the individual sub-units and the total unit are the same 'length' along the input-output dimension. The constraints on procurement of input and on disposal of output that operate on the whole unit will place upper limits on the autonomy that can be given to the sub-units. The stronger these external constraints, the greater the co-dependence of the sub-units.

Some of the problems that arise when territorially differentiated sub-units have had to be created only because of the size of the total command have been discussed in Part II. However, in other cases, so long as the territorial boundaries conform to the reality of the task structure and so long as sub-unit performance can be measured separately, this is one of the easiest kinds of command to manage, especially if the sub-units are roughly equal in size. Because the operations of his subordinates are not interdependent, the superior is not concerned with maintaining collaborative relations between them. Indeed, competitive relations are often more appropriate. Their homogeneity makes comparisons straightforward and a highly productive sub-unit can be used as an example and pace-setter for the others. Subject to the external constraints on autonomy, substantial delegation is possible, which means that a fairly large number of units can be included in one command, producing a flat hierarchy.

One practical difficulty that sometimes arises in such a command, however, is that the competitive situation gets out of hand. The superior may become so involved in resolving problems of real or imagined incomparability between the subordinates that he loses sight of the primary task of the system. The subordinates for their part are liable to seek short-term competitive advantages that may be detrimental in the long run; or alternatively they may go into collusion to protect themselves from competitive stress by establishing safely attainable norms. The common restrictive practices in industry and commerce are special cases of this form of organization.

There is another management problem that may occur in manufacturing units. This is the tendency for the sub-units to develop an 'individuality' that is based on more than their territorial differentiation from one another. Here we are not concerned with the general tendency of groups to develop a structure and culture that apparently transcend what is needed for attainment of their overt goals. We are concerned more specifically with a tendency to supplement territorial differentiation by technological differentiation. This is pace-setting of a special kind. In a manufacturing operation such as weaving, identical machinery may be used to turn out several varieties of one product. Even though all varieties are spread equitably among all sub-units, individual sub-units may develop a special proficiency in some. They acquire what Selznick has called a 'distinctive competence' (12). This distinctive competence may be encouraged, perhaps almost accidentally, by assigning more of these varieties to the sub-units in question. Such specialization is the beginning of technological differentiation. Management needs to be alert to such incipient changes and to recognize their implications. It is not simply a question of deciding whether the gains from specialization—probably in improved efficiency and quality— outweigh the disadvantages of reduced flexibility in production planning. Different methods of management are required: competition ceases to be an appropriate control mechanism when the sub-units become heterogeneous. In the extreme situation, the varieties, by ceasing to be interchangeable, acquire the status of separate products and the territorial differentiation becomes secondary to what is, in effect, technological differentiation between product units. Management of such units is discussed in the next section.

Differentiation by Technology

In cases of differentiation by technology, the notion of distinctive competence is very much present. The organization is built up around clusters of specialized skills and often specialized equipment too. Members of a sub-unit that is differentiated from others along the technological dimension derive solidarity from their distinctive competence, often by exaggerating its distinctiveness. Preservation of that distinctiveness may become the primary task of the sub-unit. Management of a unit in which the sub-units are differentiated, and therefore have to be reintegrated, along the technological dimension, involves using the specialized contributions of the sub-units to perform the primary task of the whole. To achieve this, the solidarity that the sub-units derive from distinctive skills should be sufficient for them to maintain their viability as separate systems, but not so great that they lose sight of the total task of the larger unit. To strike such a balance is no easy task. Perceived threats to the integrity and distinctiveness of sub-unit skills mobilize the energies of sub-unit members towards preservation of the sub-unit at the expense of the unit as a whole. Closed-shop movements in departments of automobile factories and demarcation disputes in shipyards are familiar examples.

Operating systems are seldom differentiated from one another by technology alone. Perhaps the nearest approximation is in enterprises such as ship-building where what is being made is also the territory of task performance. Even in ship-building, however, there is some supplementary differentiation by territory and time: certain jobs have to be done elsewhere in the yard and certain jobs on the ship itself cannot be started until preceding jobs are complete. The occupational groups at work on the ship at any one time have shifting and overlapping territorial boundaries and it is along the technological dimension that they have primarily to be co-ordinated. Conventional longwall coal-getting involves differentiation by both territory and time (5, 13, 14). The team working on a particular section of the coal-face over a 24-hour period is sub-divided into shifts that are distinguished from one another both by the times they work and by the kinds of task they do. Reinforcing differentiation by technology, territory, and time may occur in a milk business: milk is collected from the farms in the afternoon and evening and brought to the central depot; there it is processed and bottled during the night; and next morning it goes out on the delivery rounds.

In industry, technological differentiation is commonly accompanied by territorial differentiation. (The word 'department', for example, often carries both connotations.) Where the two are combined in this way, the former distinction always seems to be primary: territorial differentiation supplements and reinforces the technological. To some extent the combination also facilitates co-ordination by giving the technological groupings the security of a clear-cut physical boundary—contrasted with the vague and shifting boundaries of the shipyard.

Differentiation by Time

In the ordinary multi-shift situation, where the sub-units are differentiated from one another only by time and share a common territory and a common technology, their co-dependence is considerable. For example, maintenance failures on one shift affect the others. Generally this co-dependence is accompanied by a circular form of successional interdependence: each shift not only completes certain operations, exporting the material outside the total unit, but also passes on some semi-finished material to the next shift for completion. Throughput time is a major determinant of interdependence. The longer the throughput time, the higher the proportion of semi-finished to finished material at the end of each shift. Also, the less likely it is that individual shift performance, in terms of quantity and quality, can be measured precisely. Continuous operations of process industries provide an obvious example, but the production lines of the engineering industry also contain at any one time components in various stages of completion. Another factor that reduces the clear-cut self-containment of the shifts in the most highly automated industries, where shift working is most prevalent, is that the functions of so-called 'production workers' have increasingly been taken over by the machines themselves. The task of the workers is to monitor and maintain and the consequences of things they do or fail to do are often not immediately and clearly visible: the benefits or otherwise may fall upon succeeding shifts.

Furthermore, in most industries—indeed in the society at large—night-work is considered unnatural; a certain stigma attaches to it. Men who work while the rest of the world is asleep tend to feel cut off from society—and no doubt some select night-work for this reason, and may even become neurotically addicted to it. This is not the place for a discussion of the psychology of shift work: the point to be emphasized is that night-shifts often have a distinctive 'atmosphere' of their own.[7] This is particularly true where a group of workers is permanently on night-shift. Night-shifts are less differentiated in this particular respect in enterprises such as chemical plants, steel plants, or power stations, where continuous operations are dictated by the basic nature of the technology, and also where all shifts rotate.

It is clear that differentiation by time calls for positive managing skills to maintain the tempo and quality of work and to prevent the circular dependence from becoming a deteriorating cycle. The management problems inherent in this model make it important to eliminate

---

[7]Often, too, the level of attention is lower and mistakes are more numerous. Accidents may be fewer: cf. Hill and Trist (6).

avoidable complexities. Many of these stem from a failure on the part of management to conceptualize second and third shifts as discrete systems. Outside industries where continuous operations are intrinsic to the technology, the second and third shifts have generally been introduced in order to supplement production from single-shift working without increasing capital investment; and the notion that they are supplementary tends to be perpetuated not only in management attitudes but also in organization. Rice has given a good example of this kind of situation in a textile mill[8] and also indicated that acceptance of the organizational consequences of three-shift working can lead to higher productivity and improved quality (10).

An avoidable complication occurs, for example, when the overall head of the three shifts has himself the additional role of first-shift supervisor. The twenty-four-hour responsibility of the overall head naturally cannot be discharged if he is regularly tied for eight hours to one shift only. A separate first-shift supervisor is therefore necessary. Related to this is a tendency to confuse first-shift control and service functions with headquarters functions, usually because office hours more nearly coincide with first-shift hours than with those of other shifts. The first-shift supervisor may be given responsibility for such functions as pertain to all shifts, or alternatively—and less frequently—certain services that are decentralized to the second- and third-shift supervisors are, for the first shift, retained under headquarters control. It is appropriate either to centralize such functions fully under the head of the total command or to decentralize them equally to his three subordinates, but not to delegate them to one or two subordinates only (cf. 10, p. 46). Difficulties of co-ordination are also increased if one shift supervisor—commonly the third shift—has an operating command that is smaller than the other two. Equalization of shift commands, by allowing the heads of the three shifts to collaborate as equals, may reduce the load on the managing system to an extent that more than offsets the cost of increased third-shift working. (This is not possible, of course, where there are wide fluctuations in the load—for example, in some engineering firms—and a 'spill-over' night-shift is required irregularly in order to absorb these fluctuations and to maintain a steady day-shift load.)

The head of this kind of command therefore has to take specific precautions appropriate to the pattern of differentiation and inter-

---

[8] '. . . Supervision on the third shift had never been as good as on the other shifts. It had been unpopular with management as well as with workers. In consequence, the most junior supervisors had been posted to it. Not unnaturally the combination of least experienced workers and most junior supervisors had succeeded in proving correct those who believed that third shifts were not worth while' (10, p. 176).

dependence: he needs to be aware of his twenty-four-hour responsibility, to attend shift hand-overs as often as possible, to avoid delegating either too much or too little to the first-shift head, and to avoid giving too small a command to the third-shift head. Meetings of the superior with his subordinates as a group help to emphasize the complementary contributions of the shifts to the total task of the command. Meeting the subordinates only individually makes it more difficult to ensure that all three shifts work together coherently. There are possibly advantages in a form of shift rotation which periodically alters the order of dependence of the shifts.

Where there are only two shifts, although the general problems are very much the same as in the three-shift situation—especially the sharing of territory and equipment—the reciprocity makes equilibrium easier to sustain because dependence and power balance each other. There is one drawback in having only two shift heads reporting to one superior: it is too small a command. Co-ordination and control of two subordinates generally give the superior too little to do. He may tend to bypass his immediate subordinates, withdrawing authority and responsibility from them. Consequently it may prove desirable to combine at the same level differentiation by time with cross-cutting differentiation by territory and/or technology. As was pointed out in Part II, however, it is unlikely that the task structure will be such that interrelatedness along the time dimension will be equal to interrelatedness along the territorial/technological dimension. In all cases of cross-cutting differentiation, where two dimensions of differentiation are compressed into one level, formation of sub-groups is to be expected along one dimension or the other. It has to be realized, however, that such groupings have no formal identity in this kind of structure, so that controls and services must be either fully centralized or else fully decentralized to the individual sub-units.

Though the few models discussed here only touch the fringe of all the possible variations, they serve to indicate the different kinds of demand placed on management according to the types of boundary that separate the sub-units and according to the type and degree of dependence between them. Consideration of these factors may be relevant to the selection, training, and placement of managers. Though it is probably a little far-fetched to suggest that management of territorially differentiated units requires a special kind of person, it is certainly clear that techniques of management appropriate in that situation cannot effectively be transplanted into a situation where the units are differentiated along other dimensions and the patterns of co-dependence and interdependence are more complex.

SUMMARY

Any production system can be defined along the three dimensions of territory, technology, and time, which are intrinsic to the structure of the task of the system. Task performance is impaired if sub-systems within a system are differentiated along any other dimensions than these.

In a simple production system, management is inherent in internal relationships. A prerequisite of transition from a simple to a complex system, with a differentiated managing system, is the emergence of sub-systems with discrete sub-tasks. Simple systems vary in their 'resilience'—the capacity to resist forces towards transformation into a complex system, while remaining viable and efficient. The resilience of a simple system is increased if there is mobility of individuals between sub-systems associated with sub-tasks and if the sub-systems themselves are closely interdependent. For any expanding or changing simple system, however, there is an optimum stage for transition to a complex system: task performance suffers if the transition is premature or belated.

A large, complex production system, such as a multi-shift manufacturing concern, is broken down into progressively smaller systems along all three dimensions of territory, technology, and time. There are several levels of differentiation. Differentiation along some dimensions occurs at more than one level. Differentiation along more than one dimension can also occur at a single level. The dimensions may then be either at right angles to one another ('cross-cutting'), in which case each component sub-system is interdependent with—and has to be co-ordinated with—some sub-systems along one dimension and some along the other; or the dimensions may be coincident and mutually reinforcing (for example, departments in a factory which are differentiated from one another technologically and also occupy separate territories).

In such a complex system there is sometimes a choice: differentiation along one dimension could occur at a higher level than differentiation along another, or at a lower level. However, the basis of differentiation must not violate the task structure: boundaries should be so located as to associate each command with a 'whole' task, for which the head of the command can be held accountable. If at any level the formal organization is such that task relationships which cross the boundaries of a command are more intensive than relationships within it, there will be a loss of efficiency and/or an expansion of the managing system at that level.

The appropriate level of differentiation by time is related to the throughput time of the process (the interval between input and output).

Time differentiation occurs at a low level in an organization with a long throughput time and at a high level when throughput time is short. When a sequence of operations is involved, consecutive operations that are also territorially adjacent are appropriately differentiated at a low level, even though technologically they may be heterogeneous. Operations that are technologically homogeneous but spatially and sequentially separated can seldom be combined into a viable system.

The need to ensure that sub-commands within a command are approximately equal in size and that each has a 'whole' task may entail combining more than one system into one command or adopting mixed methods of differentiation. These needs, however, sometimes conflict.

Differentiation implies reintegration, to ensure that the sub-tasks of sub-systems add up to the total task of the whole system. Within any command, the way in which the task is broken down—i.e. the dimensions along which the constituent systems are differentiated and the interdependence between the systems—largely determines the kind of management required. Organizational models drawn from industry are used to illustrate this point.

REFERENCES

1. EMERY, F. E. *Characteristics of Socio-Technical Systems*. Tavistock Institute of Human Relations, Document No. 527, January, 1959 (mimeo).
2. GULICK, L. 'Notes on the Theory of Organization, with special reference to Government in the United States.' In *Papers on the Science of Administration*, edited by L. Gulick and L. Urwick. New York: Institute of Public Administration, 1937.
3. HERBST, P. G. 'Measurement of Behaviour Structures by Means of Input-Output Data.' *Hum. Relat.*, Vol X, No. 4, 1957, pp. 335-46.
4. HERBST, P. G. 'Task Structure and Work Relations.' T.I.H.R., Document No. 528, January, 1959 (mimeo).
5. HIGGIN, G. W. 'Studies in Work Organization at the Coal Face—I.' *Hum. Relat.*, forthcoming.
6. HILL, J. M. M., and TRIST, E. L. 'Changes in Accidents and other Absences with Length of Service.' *Hum. Relat.*, Vol. VIII, No. 2, 1955, pp. 121-52.
7. JAQUES, E. *Measurement of Responsibility*. London: Tavistock Publications Ltd., 1956; Cambridge, Mass.: Harvard University Press.
8. LEWIN, K. *Principles of Topological Psychology*. New York and London: McGraw-Hill Book Company, 1936.
9. MARCH, J. G., and SIMON, H. A. *Organizations*. New York: John Wiley & Sons, London: Chapman and Hall Ltd., 1958.
10. RICE, A. K. *Productivity and Social Organization: The Ahmedabad Experiment*. London: Tavistock Publications, 1958.
11. RICE, A. K., and TRIST, E. L. 'Institutional and Sub-Institutional Determinants of Change in Labour Turnover.' *Hum. Relat.*, Vol. V, No. 4, 1952, pp. 347-71.

12. SELZNICK, P. *Leadership in Administration: A Sociological Interpretation*. Evanston, Ill.: Row, Peterson and Company, 1957.
13. TRIST, E. L., and BAMFORTH, K. W. 'Some Social and Psychological Consequences of the Longwall Method of Coal-Getting.' *Hum. Relat.*, Vol. IV, No. 1, 1951, pp. 3-38.
14. TRIST, E. L , and MURRAY, H. 'Work Organization at the Coal Face: a Comparative Study of Mining Systems.' Forthcoming.
15. WOODWARD, J. 'Management and Technology.' Dept. of Scientific and Industrial Research: Problems of Progress in Industry, No. 3. London: H.M.S.O., 1958.

Analytical Model for Sociotechnical Systems

Fred E. Emery and Eric L. Trist

PURPOSE OF MODEL

The analytical model has been developed as a practical tool to help line managers implement the concept of joint optimization in their own departments or sections. It is hoped that it will enable managers to examine their existing technical systems and their existing organizations to gain insight into the technical and social systems, and to improve the level of performance.

The model is concerned solely with the *analysis* of production systems and the development of change proposals. Clearly, in practice there would need to be a preceding step of selecting an appropriate area for analysis and allocating people to carry it out and a succeeding implementation phase, in which the change proposals would be developed into an action program, changes would be initiated, and some sort of evaluation mechanism would be set up.

THE MODEL IN PRACTICE

As the model has been tried out on a number of different production systems, certain lessons have emerged:

1. Care must be taken in selecting the appropriate area for analysis. Existing organizational boundaries are not neccessarily the most appropriate ones. In practice, it is all right to select a production system as defined by current organizational boundaries, bearing in mind that one result of the analysis might be an indication that the boundaries should be adjusted. Departmental management should not

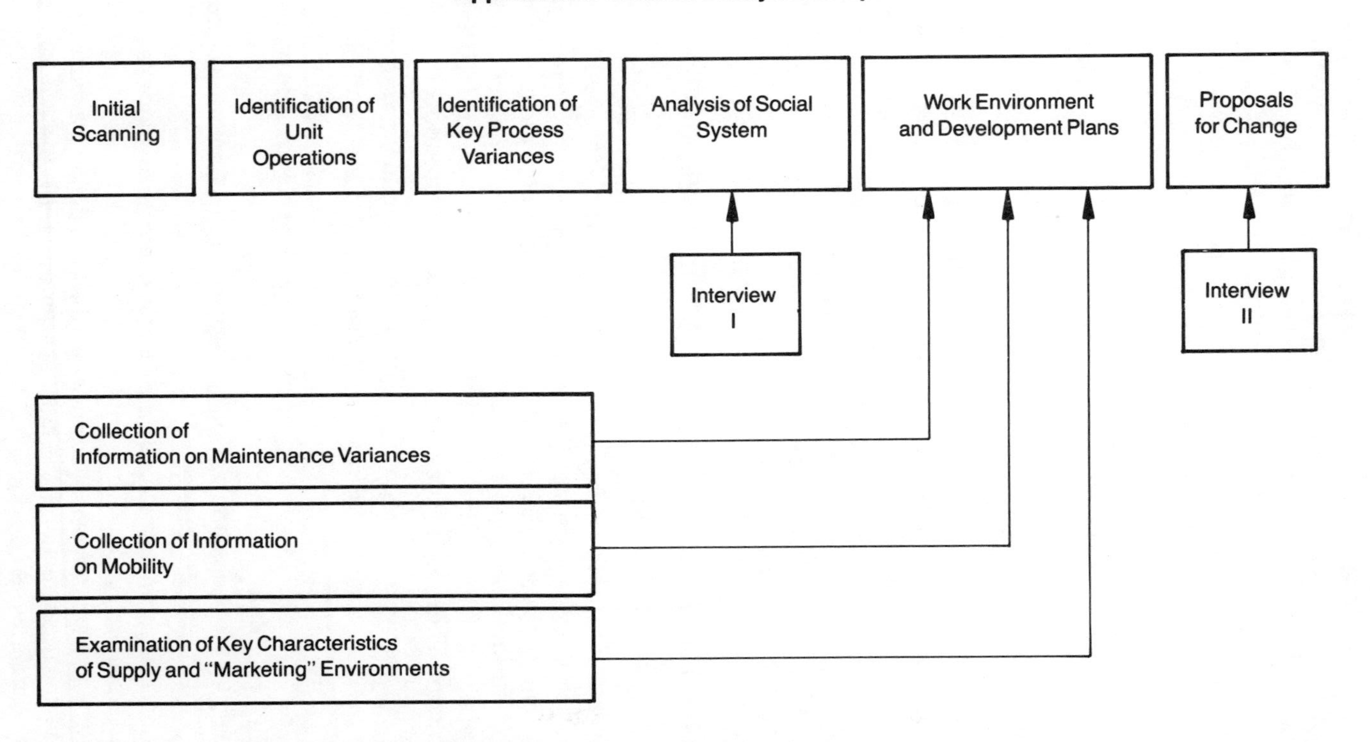

Figure 1. Analytical Model for Sociotechnical Systems

select too large or too complex an area for analysis, at least initially. In most cases, a department is too large a unit for full study within a reasonable period of time, with the resources available. Greater progress can be made by selecting smaller units, if possible.

2. For analysis, it is helpful to concentrate on the production system as it is currently operating; otherwise, some confusion may creep in with reference to the way it "used" to operate or how it "might" operate in the future.

3. A practical difficulty is the tendency to collect too much detail. An effort should be made to identify only key information under each step heading and to avoid getting caught up with an abundance of detail.

4. To initiate and carry through an analytical process and to carry out an ensuing action program will require a high degree of effort and commitment. It will also require the participation of people at all levels in the production systems concerned. It seems highly desirable, therefore, to set up a small action group headed by the production system manager, with representatives from various levels and with such outside help as may be required.

ANALYTICAL MODEL FOR SOCIOTECHNICAL SYSTEMS

Step 1: Initial Scanning

The objectives of this step are to identify the main characteristics of the production system and of the environment in which it exists and to determine, if possible, the main problems and where the main emphasis of the analysis needs to be placed. This can be done through a carefully structured briefing of the action group by a departmental manager. The briefing should cover the following ground:

a. The general geographical layout of the production system.
b. The existing organizational structure and the main groupings within it.
c. The main inputs into the system, with specifications where appropriate.
d. The main outputs from the system, again with specifications where appropriate.
e. The main transforming processes that take place within the system.
f. The main types of variance in the production system and their sources, e.g., the nature of raw material, the equipment, or breakdowns.

g. The main characteristics of the relationship between the production system and the department in which it exists.
h. The objectives of the system, both production and social.

Step 2: Identification of Unit Operations

The purpose of this step is to identify the main phases in the production operation. Unit operations are the main segments or phases in the series of operations that have to be carried out to convert materials at the input end of the system into products at the output end. Each unit operation is relatively self-contained, and each effects an identifiable transformation in the raw material; a transformation in this sense is either a change of state in the raw material or a change of location or storage of the material.

The actions necessary to effect the transformation may be carried out by machines or by people, but we are not concerned at this stage with either the characteristics or needs of the machines (e.g., maintenance needs or operating characteristics) or the characteristics and needs of the people (e.g., psychological needs). Attention is entirely on the series of transformations through which the raw material goes. Where possible, the purpose of each unit operation needs to be identified in terms of its inputs, its transformations, and its outputs.

Step 3: Identification of Key Process Variances and Their Interrelationships

The objectives of this stage are to identify the key process variances and the interrelationships between them. A variance is a deviation from some standard or from some specification.

It is necessary to emphasize that, at this stage in the analysis, we are concerned with variance that comes from the raw material or from the nature of the process itself as it is currently or normally operating. We are not concerned with variance that comes from faults in the technical equipment or plant (e.g., breakdown or malfunctioning) nor are we concerned with variance that comes from the social system (e.g., maloperation or human error).

We also are not concerned with the total range of variance. From other studies it has been found that there are a large number of variances in any production system that have either no effect or a comparatively minor effect on the ability of the production system to pursue its objectives. It may be necessary, however, to take some such variance into account in subsequent attempts to reach a higher level of joint optimization, but at this stage we are concerned only with those "key" variances that significantly affect the capability of the production system to pursue its objectives in one or more of its unit operations. The sequence of actions necessary is as follows:

a. Identification of all variances in the system (arising from the nature of the raw material or from the nature of the process) that, in the opinion of the action group, are worthy of note. The main sources of information are the manager and supervisors of the system, who draw on their knowledge and experience. It is necessary to go through the process of identifying variances several times to ensure that all the main variances have been included.
b. Drawing up a matrix of the variances identified. This shows any clusters of variances—control problems—and also shows where information loops exist or are necessary in the production system. It will also help in the selection of key variances (e.g., variances that have an effect through a series of unit operations are likely to be considered key to the control of the process).
c. The identification of the key variances can be done in two stages:
   (1) The department or unit manager and appropriate assistants should make out a list of what they consider to be the key variances, drawing on their experience and knowledge of the production system;
   (2) The action group should work with this list, checking it against the matrix of variances and against the following four criteria. A variance should be considered "key" if it significantly affects:
      - quantity of production
      - quality of production
      - operating costs (use of utilities, raw material, overtime, etc.)
      - social costs (e.g., the stress, effort, or hazard imposed on the employees).

The first three dimensions are concerned with the system's *production objectives*. The last is concerned with the production system's *social objectives*. It is possible to move now to an analysis of the social system, to examine the way in which it contributes to control of variances and so to the attainment of the production system's objectives, and to examine the extent to which the social system's own needs are met.

Step 4: Analysis of the Social System

The objective of this step is to identify the main characteristics of the *existing* social system. Its complex sets of interrelations and groupings, both formal and informal, do not need to be described. By structuring the analysis carefully, it is possible for the analytical team to draw out sufficient of the relevant information to enable it to begin to develop job-design proposals relatively quickly. The following steps are the minimum necessary:

a. A *brief review* of the organizational structure where necessary, filling in a little more detail than was included in Step 1 on number of levels, social groupings, and types of roles.
b. A *table of variance control* to show the extent to which key variances are presently controlled by the social system. It is possible to identify where key organizational and informational loops exist or are required by using such a table.

It answers the following questions: Where in the process does the variance occur? Where is it observed? Where is it controlled and by whom? What tasks must the controller do to control it? and What information does the controller have from what source to carry out these control activities? Hypotheses that are formed should be noted for subsequent discussion and possible validation.

c. *Ancillary activities* such as descriptions of the workers' roles in the production system should be noted. Activities connected with the control of key variances will be listed in the variance control table. It is likely, however, that there will be a number of ancillary activities. Identifying these and trying to relate them to the control process may well lead to the identification of additional key variances. On the other hand, it could conceivably lead to the elimination of these ancillary activities altogether.
d. *Spatial and temporal relationships* such as the physical or geographical relationships between the various roles in the production system (e.g., distance or physical barriers between workers) and their relationship over time must be mapped out.
e. *Flexibility*, i.e., the extent to which the workers share a knowledge of each other's roles, can be identified on a mobility chart. It may be necessary to carry out this step in two phases: an initial analysis of rotation and a more detailed analysis of the extent to which they carry out the essential tasks associated with the roles. The chart should cover a period of two or three months. Therefore, it is good to start recording this information in the early weeks of the analysis process.

f. *The payment system* and how it is related to various roles in the production system must be studied because it has an impact on job rotation, group working, etc.
g. *The psychological needs* of employees can be tested to see if their roles meet these needs. An adequate/inadequate rating for each employee's main activities is sufficient. For this purpose, the action group will need to rely on management perceptions of the roles. To learn the workers' perceptions of their roles, it will be necessary to set up some machinery for the collection of their views.
h. *Identify areas of maloperation* to establish causes, where possible.

Step 5: People's Perceptions of Their Roles

This step, although it is part of an analysis of the social system, is dealt with separately partly because of its importance and partly because of the method of carrying it out. Its purpose is to learn as much as possible of the people's perceptions of their roles, *specifically the extent to which they see them fulfilling their psychological needs.* This can be accomplished by a personnel worker in the action group, either for this particular purpose or as a full member.

Two interviews can be arranged with appropriate groups of workers (1) within the first six weeks of the analysis and (2) toward the end of the process, when job-design proposals are being finalized. Both interviews must be highly structured, designed with open-ended questions based on psychological needs and, in the case of the second interview, on the developing job-needs design proposals.

With this step, the analysis of the production system itself is complete, and it is to be expected that a number of redesign proposals or hypotheses will have emerged.

The analysis now considers the impact on the production system of a number of "external" systems, e.g., maintenance, supply and user systems, personnel policy, etc., that will influence any hypotheses that have emerged and may well bring out further redesign proposals.

Step 6: Maintenance System

This step is *not* concerned with the examination of the maintenance system or organization as such, but solely with the extent to which that system has an impact on the particular production system being analyzed and with the extent to which the maintenance system affects the capability of the production system to achieve its objectives.

These objectives are:

a. To determine the nature of the maintenance variance arising in the production system.
b. To determine the extent to which that variance is controlled.
c. To determine the extent to which maintenance tasks should be taken into account in the design of operating roles.

This analysis of maintenance variance is not in any way subordinate to the analysis of process variance carried out in Step 3. Both are necessary to an understanding of the characteristics of the production system. It may be that in some cases variance of a greater order comes from the maintenance system than from the production system itself, in which case one would expect greater emphasis on this stage.

To collect information on maintenance activities, the analyst must begin with the first month of the project and continue for two or three months. The collection of additional data and the burden of collection placed on operating and maintenance staff should be kept to the minimum consonant with achieving the objectives of the analysis.

Step 7: Supply and User Systems

This step is *not* concerned with identifying the characteristics of the supply and user systems themselves, but with the way in which these environmental systems affect the particular production system. The objectives of this stage are:

a. To identify the variances that are passed into the production system from the supply and user systems.
b. To examine, where appropriate, the extent to which these variances could be controlled closer to their source or their effect on the production system could be diminished.

In general, the analysis across the boundaries of the production system should be kept at a fairly general level initially and should go into greater detail only if there appears to be a real possibility of effecting an improvement, e.g., a better control of variance or more appropriate flow of information.

The result of this step might either be a diminishing of the variance arising in the production system from across its boundaries or, in some cases, *a redefining of the production system's objectives* to ensure that they realistically take into account both supply and "marketing" constraints.

Step 8: Work Environment and Development Plans

The purpose of this step is to identify those forces operating in the wider departmental or work environment that either affect the production system's ability to *achieve* its objectives or that are likely to lead to a *change* in those objectives in the foreseeable future. It has two main steps:

a. *Development Plans*. The identification of any plans, either short or long term, that have a high probability of being implemented *for the development of the social or the technical systems*. These clearly would have to be taken into account in the development of any redesign proposals.

b. *General Policies*. The identification of any general policies or practices that impinge on the production system, if these have not been taken into account in the examination of the maintenance system and the supply/user systems. Examples are the general method of promotion, which affects the social system, or the utilities supply and control system operating throughout the plant, which affects the technical system.

It should be emphasized that we are not concerned with the characteristics of these environmental systems as they exist in themselves, but only insofar as they affect the ability of the production system to pursue its objectives. In the analysis of most production systems, these environmental factors will constitute "givens" rather than areas to be included in proposals for change.

Step 9: Proposals for Change

The purpose of this step is to gather all the hypotheses and proposals that have been developed during the analysis process, to consider their viability, and to make them the basis of a subsequent action program.

As was mentioned earlier, it is likely that hypotheses will be formed as the analysis of the technical system is being completed. These probably will be expanded, eliminated, or modified as further information is gathered about the social and environmental systems.

Those hypotheses that remain must be tested, as much as is possible on a theoretical basis, against appropriate criteria before being developed into viable proposals. The actual criteria will vary and will require careful design, but these criteria must relate to the production system objectives and must cover:

a. The production objectives of the system, i.e., in terms of quantity, quality, and general operating costs. This covers proposals specifically aimed at increasing the control over or diminishing variance in the production system.

b. The social objectives of the production system, such as those aimed at increasing the extent to which psychological needs are met in role design or diminishing the costs borne by the work force (e.g., stress, hazard, or heavy labor).

Many proposals will, of course, lie in both areas. For example, proposals aimed at increasing the level of responsibility at the lower levels would meet psychological requirements and might lead to shorter lines of communication and more effective variance control. In addition, any proposals for the redesign of the social system must be tested against emergency and crisis needs. In the case of a process unit, this would entail the ability to shut the unit down in the event of a loss of power or feed or a major fire.

ORGANIZATIONAL OBJECTIVES AND ROLE ANALYSIS

This model has been developed in conjunction with the sociotechnical model as an alternative method of analysis for departments in which no continuous process exists, e.g., service or advisory departments. Like the sociotechnical model, its purpose is to help managers analyze their existing organizations, as they currently and normally operate, and to develop proposals for change when this seems likely to lead to improved performance. The model is still in the development stage.

Step 1: General Scanning

This should provide a general introduction to the outputs, inputs, and transformation processes in the department, i.e., its objectives, its work, and its organizational structure and location within the organization, as well as the geographical layout of the department. Scanning is necessary so that more detailed investigations at a later time can be seen against an overall background. In general, the amount of detail collected should be kept small. It is probably useful for the departmental manager, if this analysis is undertaken by an action group, to describe each process itself rather than to explain its purpose.

Step 2: The Objectives of the System

It is important to arrive at a clear definition of objectives because this provides a rational datum against which to judge all activities in the department. In practice, the identification and statement of objectives pose difficulties because:

a. The objectives stated may be too general;
b. They may be multiple, but only one or two may be identified;
c. They may be nonmeasureable;
d. They may refer to several time periods;
e. They may be partly derived from higher or other system levels;
f. They may be outputs that the system wants to minimize rather than to maximize, e.g., waste;
g. They may involve changes of the internal structure of the system rather than outputs from the system, e.g., change in assets;
h. They may not be well enough recognized to be formulated.

To cope with these problems, the following method of analysis is proposed:

Consider all major *outputs* of the department, whether processed raw materials, communications, or workers. Then try to identify all *inputs* and follow these inputs to determine the steps they go through to become outputs. Make sure that no significant output has been missed. Then test outputs to determine whether they are objectives by presenting them to the manager of the next higher system level and asking him or her whether these are indeed the outputs required. Also, some inputs come into the department to maintain or develop the assets, and part of the objectives of a department will be directed toward these two activities. The assets of a department include its plant and equipment, the money over which the manager may have authority, and the workers.

When considering a department's outputs, a problem may arise because it may not be possible to describe an output such as a communication (written or verbal) meaningfully unless some indication is given of its necessary contribution to an overall decision that is made outside the boundaries of the department. In such cases, it is useful to draw up a table with the following headings:

- Description of output
- To whom sent
- Overall decision it was intended to support
- Required contribution of the department's output to this decision
- Consequences of substandard performance on the social and economic cost (both inside and outside the department)

This analysis will determine the resources that are within the boundaries of the department and those the manager needs to bring in. Because departmental objectives are more clear, it should be possible to hypothesize the responsibilities, authorities, information/communication links with others, and key methods and procedures that are appropriate and to match them against those that already exist.

Step 3: Analysis of the Roles in the System

An analysis must be done also of each role in the system to arrive at the role objectives and to relate them to the overall departmental objectives. This process should start with the manager's role and work down.

Step 4: Grouping of Roles

This analysis will identify the necessary role-interaction links insofar as the current process exists and will lead to hypotheses about the clustering of these roles in respect to their geographical and temporal distribution and status dimensions.

Step 5: Measurement of Roles Against Psychological Requirements

After identifying the inputs, transformations, and outputs of each role, it is useful to measure the manager's perception and the workers' own perceptions of how much each role meets the workers' psychological requirements. The workers' perceptions of their own roles can be learned through individual interviews, preferably conducted by someone outside the department.

Step 6: Develop Change Proposals

In the course of the preceding steps, various hypotheses for change will have emerged. These should be refined into proposals for the redesign of jobs or organizations, e.g., a change in authorities or methods of grouping, or it may be that analysis by this stage will have indicated a need for a reformulation of departmental objectives. Proposals for change will, of course, have to be related to the overall environment of a department.

Step 7: Management by Objectives

Once the objectives of the department and its constituent roles have been determined, attention should be given to developing performance measures, to setting targets (either jointly with a manager or by oneself), and to feeding back these results to the person occupying the role. However, because important areas of role output may not be measured readily, care should be taken to ensure that these areas are in some way included in performance targets so that role output is not distorted.

PART IV-B:

SOCIOTECHNICAL CHANGE

Introduction

After the sociotechnical system has been analyzed using the guidelines set forth in the preceding articles, the next task is to proceed with actual sociotechnical system change. The following articles on system change vary a great deal in content. Some are theoretical; others provide outlines for change; and still others describe actual efforts at change. We believe that such variety is consistent with the complexities of change processes in organizations and provides the balanced perspective necessary for understanding sociotechnical change.

Because sociotechnical methods deal both with people and with technologies, it is important that the change agent know something of both. Simply to recognize the need for change is insufficient if that recognition is not shared and accepted by those affected and if the direction of the change process is not clear to those involved. The articles in this section provide insight into the interaction between social and technological systems and clues about how specific changes will affect this interaction. With the help of this information, the practitioner can undertake sociotechnical interventions in organizations, realizing that each organization is unique and that thoughtful adaptations of the suggestions provided here are required.

Beckhard presents a general theory of intervention in organizations and accounts of his efforts for change in large-scale systems. He provides a model for the planning of change that is directly applicable to sociotechnical efforts. It includes defining the change problem, determining readiness and capability for change, identifying the consultant's own resources and motivation for change, and determining the immediate change strategy and goals. Examples of large-system interventions are given, including efforts to change the relationship of the organization to its environment, to change managerial strategy, to change organizational structures, to change the way work is done, and to change reward systems. Each intervention may be suited to reinforcing sociotechnical change in a particular organization. Suggestions and alternatives are offered concerning early intervention possibilities and methods for maintaining changes once they have been introduced.

Davis' paper, "The Coming Crisis for Production Management," focuses on the inadequacy of traditional organizational structures for dealing with changes occurring in technology, society, and the economy. Davis contends that organizations must respond to these changes with new sociotechnically designed structures that incorporate responsible autonomy, adaptability, variety, and participation.

Cummings provides a detailed, practical guide to sociotechnical intervention. His outline includes: defining the experimental system; sanctioning and experimenting; forming an action group; analyzing the experimental system; generating hypotheses for redesign; testing and evaluating hypotheses for redesign; providing transition to a normal operating system; and disseminating the results. The article also includes a presentation of work-redesign principles originally set forth by Emery. These include suggested sociotechnical changes at the individual, group, and organizational levels. Cummings' article and the articles by Emery and Trist and Beckhard in this section provide a basic outline that can be used in the majority of sociotechnical change efforts.

The two articles by Walton include detailed accounts of sociotechnical interventions, illustrating what an effectively functioning, sociotechnically designed organization "should" look like. In the first article, Walton outlines several causes of worker alienation and describes a sociotechnical system design effort in the General Foods Corporation, a prototype of change to overcome that alienation.

In the second article, Walton describes the characteristics of sociotechnical interventions in several countries. As Beckhard has done, Walton discusses problems of survival and growth of sociotechnical interventions, concluding that:

> even after successful introduction and early signs of effectiveness, a host of other factors threaten the continued viability of the redesigned units and frustrate efforts to extend innovations to other units. (p. 234)

If only to learn the effort and resources that must be expended to produce successful sociotechnical change, Walton's discussion of the maintenance of change should be required reading for the sociotechnical practitioner.

Finally, excerpts of various sociotechnical projects undertaken by the Philips organization are included as additional examples of the effects of specific sociotechnical changes in a variety of settings. As the author states:

> The examples chosen are not, however, intended as *ideal* projects or as being *representative* of a particular country or technology. The main reason for aiming at the greatest possible diversity was to make it easier for others to recognize situations and problems similar to their own and thus to stimulate them to similar activities. (p. 237; italics in original)

Together the articles on sociotechnical diagnosis and change give a foundation for successful innovation. Although we know much more today than we did ten or twenty years ago, our understanding of organizational change is still incomplete; successful efforts at change are never guaranteed, even for the most experienced practitioner.

Strategies for Large System Change*

Richard Beckhard

INTRODUCTION

The focus of this article is on assisting large organization change through consultative or training interventions. As used below, "client" refers to an organization's leader(s) and "consultant" refers to the intervenor or change facilitator. Note that the consultant can come from within or from outside the organization.

Intervention is defined here as behavior which affects the *ongoing social processes* of a system. These processes include:

1. Interaction between individuals.
2. Interaction between groups.
3. The procedures used for transmitting information, making decisions, planning actions, and setting goals.
4. The strategies and policies guiding the system, the norms, or the unwritten ground rules or values of the system.
5. The attitudes of people toward work, the organization, authority, and social values.
6. The distribution of effort within the system.

Interventions can affect any one or several of these processes.

---

*This article is adapted from a chapter by the author in *Laboratory Method of Changing and Learning*, Benne, Bradford, Gibb, and Lippitt, editors. Science and Behavior Books. Palo Alto, California, 1975.

Reprinted from Richard Beckhard, "Strategies for Large System Change," SLOAN MANAGEMENT REVIEW, Winter 1975, pp. 43-55. Used with permission.

The first part of this article describes a model of diagnosis and strategy planning which has had high utility for the author during the past several years. The second part examines a number of actual strategies in organization and large system change and the issues of where to begin change and how to maintain change.[1]

A MODEL FOR CHANGE PLANNING

The following model is far from perfect. However, its use seems to enable one to ask the "right" questions and to obtain answers that yield a basis for relatively trustworthy judgment on early interventions into the large system. For convenience the model will be discussed under four headings.

Defining the Change Problem

When a change effort is initiated, either the client and/or the consultant, or some other part of the system has determined that there is some need for change. An initial diagnostic step concerns analyzing what these needs are and whether they are shared in different parts of the system. For example, let us suppose top management in an organization sees as a major need the improvement of the supervisory behavior of middle management and, simultaneously, the personnel staff in the organization sees as a *prior* need a change in the behavior of the top management and a change in the reward system. These are two very different perceptions of the priority of need for initial change, but a common perception that there is a need for change in the organization does exist. As a part of determining the need for change, it is also useful to collect some information from various parts of the system in order to determine the strength of the need.

There are two distinct ways of defining the change problem. The first considers the *organization* change needed or desired. For example, does the need concern changing the state of morale, the way work is done, the communication system, the reporting system, the structure or location of the decision making, the effectiveness of the top team, the relationships between levels, the way goals are set, or something else? The second considers what *type* of change is desired and what the hierarchy or rank-ordering of these types is. One should ask whether the primary initial change requires a change:

---

[1]For a more detailed explanation of the author's views concerning organization development and intervention, see Beckhard [1].

1. Of attitudes? Whose?
2. Of behavior? By whom and to what?
3. Of knowledge and understanding? Where?
4. Of organization procedures? Where?
5. Of practices and ways of work?

Rank-ordering the various types of change helps to determine which early interventions are most appropriate.

Having defined the change problem or problems from the viewpoint of both organizational change and change process, one can look at the organization system and subsystems to determine which are primarily related to the particular problem. The appropriate systems may be the organizational hierarchy, may be pieces of it, may be systems both inside and outside of the formal structure, or may be some parts of the formal structure and not other parts. A conscious identification of those parts of the total system which primarily affect or are affected by the particular change helps to reduce the number of subsystems to be considered and also helps to clarify directions for early intervention.

Determining Readiness and Capability for Change

Readiness as stated here means either attitudinal or motivational energy concerning the change. Capability means the physical, financial, or organizational capacity to make the change. These are separate but interdependent variables.

In determining readiness for change, there is a formula developed by David Gleicher of Arthur D. Little that is particularly helpful. The formula can be described mathematically as $C = (abd) > x$, where C = change, a = level of dissatisfaction with the status quo, b = clear or understood desired state, d = practical first steps toward a desired state, and x = "cost" of changing. In other words, for change to be possible and for commitment to occur there has to be enough dissatisfaction with the current state of affairs to mobilize energy toward change. There also has to be some fairly clear conception of what the state of affairs would be if and when the change were successful. Of course, a desired state needs to be consistent with the values and priorities of the client system. There also needs to be some client awareness of practical first steps, or starting points, toward the desired state.

An early diagnosis by the consultant of which of these conditions does not exist, or does not exist in high strength, may provide direct clues concerning where to put early intervention energy. For example, if most of the system is not really dissatisfied with the present state of things, then early interventions may well need to aim toward increasing the level of dissatisfaction. On the other hand, there may be plenty of dissatisfaction with the present state, but no clear picture of what a desired state might be. In this case, early interventions might be aimed at getting strategic parts of the organization to define the ideal or desired state. If both of these conditions exist but practical first steps are missing, then early intervention strategy may well be to pick some subsystem, e.g., the top unit or a couple of experimental groups, and to begin improvement activities.

The following case illustrates these ideas. A general manager was concerned that the line managers were not making good use of the resources of the staff specialists. He felt that the specialists were not aggressive enough in offering their help. He had a desired state in mind of what good use of staff by line would be. He also had a practical first step in mind: send the staff out to visit the units on a systematic basis and have them report to him after their visits. The manager sent a memo to all staff and line heads announcing the plan. Staff went to the field and had a variety of experiences, mostly frustrating. The general manager got very busy on other priorities and did not hold his planned follow-up meetings. After one round of visits, the staff stopped its visits except in rare cases. Things returned to normal. An analysis showed that the general manager's real level of dissatisfaction with the previous state of affairs was not high enough to cause him to invest personal energy in follow-up reporting, so the change did not last.

Capability as defined here is frequently but not always outside of personal control. For example, a personnel or training manager may be ready to initiate a management development program but have low capability for doing it because he has no funds or support. The president of an organization may have only moderate or low readiness to start a management development program but may have very high capability because he can allocate the necessary resources. Two subordinates in an organization may be equally ready and motivated towards some change in their own functioning or leadership skills. One may have reached the ceiling of his capabilities and the other may not. Looking at this variable is an important guideline in determining interventions.

Identifying the Consultant's Own Resources and Motivations for the Change

In addition to defining the client and system status, and determining with the client the rank-ordering of change priorities, it is necessary for the consultant to be clear with himself and with the client about what knowledge and skills he brings to the problems and what knowledge and skills he does not have. One of the results of the early dependency on a consultant, particularly if the first interventions are seen as helpful or if his reputation is good, is to transfer the expertise of the consultant in a particular field to others in which his competence to help just is not there.

Concerning motivations, one of the fundamental choices that the consultant must make in intervening in any system is when to be an advocate and when to be a methodologist. The values of the consultant and the values of the system and their congruence or incongruence come together around this point. The choice of whether to work with the client, whether to try to influence the client toward the consultant's value system, or whether to take an active or passive role is a function of the decision that is made concerning advocacy vs. methodology.

This is not an absolute decision that, once made at the beginning of a relationship, holds firm throughout a change effort. Rather, it is a choice that is made daily around the multitude of interventions throughout a change effort. The choice is not always the same. It is helpful to the relationship and to the change effort if the results of the choice are known to the client as well as to the consultant.

Determining the Intermediate Change Strategy and Goals

Once change problems and change goals are defined, it is important to look at intermediate objectives if enough positive tension and energy toward change are to be maintained. For example, let us suppose that a change goal is to have all of the work teams in an organization consciously looking at their own functioning and systematically setting work priorities and improvement priorities on a regular basis. An intermediate goal might be to have developed within the various divisions or sections of the organization at least one team per unit by a certain time. These *intermediate* change goals provide a target and a measuring point en route to a larger change objective.

One other set of diagnostic questions concerns looking at the subsystems again in terms of:

1. Readiness of each system to be influenced by the consultant and/or entry client.
2. Accessibility of each of the subsystems to the consultant or entry client.
3. Linkage of each of the subsystems to the total system or organization.

To return to the earlier illustration concerning a management development program, let us suppose that the personnel director was highly vulnerable to influence by the consultant and highly accessible to the consultant but had low linkage to the organization, and that the president was much less vulnerable to influence by the consultant and the entry client, here the training manager. Then the question would be who should sign the announcement of the program to line management. The correct answer is not necessarily the president with his higher linkage nor the personnel man with his accessibility and commitment. The point is that weighing these three variables helps the consultant and client to make an operational decision based on data. Whether one uses this model or some other, the concept of systematic analysis of a change problem helps develop realistic, practical, and attainable strategy and goals.

INTERVENTION STRATEGIES IN LARGE SYSTEMS

The kinds of conditions in organizations that tend to need large system interventions will now be examined.

Change in the Relationship of the Organization to the Environment

The number and complexity of outside demands on organization leaders are increasing at a rapid pace. Environmental organizations, minorities, youth, governments, and consumers exert strong demands on the organization's effort and require organization leaders to focus on creative adaptation to these pressures. The autonomy of organizations is fast becoming a myth. Organization leaders are increasingly recognizing that the institutions they manage are truly *open* systems. Improvement strategies based on looking at the internal structure, decision making, or intergroup relationships exclusively are an incomplete method of organization diagnosis and change strategy. A more relevant method for today's environment is to start by examining how the organization and its key subsystems relate to the different environments with which the organization interfaces. One can then determine what kinds of organization structures, procedures and practices will allow each of the units in the organization to optimize the interface with its different environment. Having identified these, management can turn its energy toward the problems of integration (of standards, rewards, communications systems, etc.) which are consequences of the multiple interfaces.

The concept of differentiation and integration has been developed by Paul Lawrence and Jay Lorsch.[2] In essence, their theory states that

---

[2]See Lawrence and Lorsch [4].

within any organization there are very different types of environments and very different types of interfaces. In an industrial organization, for example, the sales department interfaces with a relatively volatile environment: the market. The production department, on the other hand, interfaces with a relatively stable environment: the technology of production. The kind of organization structure, rewards, work schedules and skills necessary to perform optimally in these two departments is very different. From a definition of what is appropriate for each of these departments, one can organize an ideal, independent structure. Only then can one look at the problems of interface and communication.

Clark, Krone and McWhinney[3] have developed a technology called "Open Systems Planning" which, when used as an intervention, helps the management of an organization to systematically sharpen its mission goals; to look objectively at its present response pattern to demands; to project the likely demand system if no pro-active actions are taken by the organization leadership; to project an "ideal" demand system; to define what activities and behavior would have to be developed for the desired state to exist; and finally to analyze the cost effectiveness of undertaking these activities. Such a planning method serves several purposes:

1. It forces systematic thinking.
2. It forces people to think from outside-in (environment to organization).
3. It forces empathy with other parts of the environment.
4. It forces the facing of today's realities.
5. It forces a systematic plan for priorities in the medium-term future.

This is one example of large system intervention dealing with the organization and its environment. Another type of intervention is a survey of organization structure, work, attitudes and environmental requirements. From this an optimum organization design is developed.

There is an increasing demand for assistance in helping organization leaders with these macro-organization issues. Much current change agent training almost ignores this market need. Major changes in training are called for if OD specialists are to stay organizationally relevant.

Change in Managerial Strategy

Another change program involving behavioral science oriented interventions is a change in the *style* of managing the human resources of

[3]See Krone [3].

the organization. This can occur when top management is changing their assumptions and/or values about people and their motivations. It can occur as a result of new inputs from the environment, such as the loss of a number of key executives or difficulty in recruiting top young people. It can occur as females in the organization demand equal treatment or as the government requires new employment practices. Whatever the causes, once such a change is planned, help is likely to be needed in:

1. Working with the top leaders.
2. Assessing middle management attitudes.
3. Unfreezing old attitudes.
4. Developing credibility down the line.
5. Dealing with interfacing organizations, unions, regulatory agencies, etc.

Help can be provided in organization diagnosis, job design, goal setting, team building, and planning. Style changes particularly need considerable time and patience since perspective is essential and is often lost by the client. Both internal and outside consultants can provide significant leadership in providing perspectives to operating management. Some of the questions about key managers that need answers in planning a change in managerial strategy are:

1. To what degree does the top management encourage influence from other parts of the organization?
2. How do they manage conflict?
3. To what degree do they locate decision making based on where information is located rather than on hierarchical roles?
4. How do they handle the rewards that they control?
5. What kind of feedback systems do they have for getting information about the state of things?

Change in the Organization Structures

One key aspect of healthy and effective organizations is that the structures, the formal ways that work is organized, follow and relate to the actual work to be done. In many organizations the structure relates to the authority system: who reports to whom. Most organizations are designed to simplify the structure in order to get clear reporting lines which define the power relationships.

As work becomes more complex, it becomes impossible in any large system to have *one* organization structure that is relevant to all of the kinds of work to be done. The basic organization chart rarely describes the way even the basic work gets done. More and more organization leaders recognize and endorse the reality that organizations actually operate through a variety of structures. In addition to the permanent organization chart, there are project organizations, task forces, and other temporary systems.

To clarify this concept, we examine a case where a firmly fixed organizational structure was a major resistance to getting the required work done. In this particular consumer-based organization there was a marketing organization that was primarily concerned with competing in the market, and a technical subsystem that was primarily concerned with getting packages designed with high quality. Market demands required that the organization get some sample packages of new products into supermarkets as sales promotions. The "rules of the game" were that for a package to be produced it had to go through a very thorough preparation including design and considerable field testing. These standards had been developed for products which were marketed extensively in markets where the company had a very high share. The problem developed around a market in which the company had a very low share and was competing desperately with a number of other strong companies. Because of the overall company rules about packages, the marketing people were unable to get the promotion packages into the stores on time. The result was the loss of an even greater share of the market. The frustration was tremendous and was felt right up to the president.

Within the marketing organization there was a very bright, technically oriented, skilled, abrasive entrepreneurial person, who kept very heavy pressure on the package technical people. He was convinced that he could produce the packages himself within a matter of weeks as opposed to the months that the technical people required. Because of his abrasiveness he produced much tension within the technical department and the tensions between the two departments also increased. At one point the heads of the two departments were on a very "cool" basis. The president of the company was quite concerned at the loss of markets. He had attempted to do something earlier about the situation by giving the marketing entrepreneur a little back room shop in which he could prove his assertions of being able to produce a package in a short time. The man did produce them, but when he took them to the technical people for reproduction, they called up all the traditional ground rules and policies to demonstrate that the package would not work and could not be used.

The client, here the president, had diagnosed the problem as one of noncooperation between departments and particularly between individuals. Based on this diagnosis he asked for some consultative help with the interpersonal problem between the marketing entrepreneur and the people in the technical department. He also thought that an intergroup intervention might be appropriate to increase collaboration between the groups.

The consultant's diagnosis was that although either of these interventions was possible and might, in fact, produce some temporary change in the sense of lowering the heat in the situation, there was little possibility of either event producing more packages. Rather, the change problem was one of an inappropriate structure for managing work.

The consultant suggested that the leaders of marketing and technical development together develop a flow chart of the steps involved in moving from an idea to a finished promotion package. Then they were to isolate those items which clearly fitted within the organization structure, such as the last few steps in the process which were handled by the buying and production department. The remaining steps, it was suggested, needed to be managed by a *temporary* organization created for just that purpose. The consultant proposed that for each new promotion a temporary management organization be set up consisting of one person from packaging, one from marketing, one from purchasing, and one from manufacturing. This organization would have, as its charter, the management of the flow of that product from idea to manufacturing. They would analyze the problem, set a timetable, set the resource requirements and control the flow of work. The resources that they needed were back in the permanent structures, of which they were also members. This task force would report weekly and jointly to the heads of both the technical and marketing departments. The president would withdraw from the problem.

The intervention produced the targeted result: promotion packages became available in one-fifth of the time previously required. The interpersonal difficulties remained for some time but gradually decreased as people were forced to collaborate in getting the job done.

Change in the Ways Work Is Done

This condition is one where there is a special effort to improve the meaningfulness as well as the efficiency of work. Job enrichment programs, work analysis programs, and development of criteria for effectiveness can all be included here. To give an example, an intervention might be to work with a management group helping them examine their recent meeting agendas in order to improve the allocation of work tasks. Specifically, one can get them to make an initial list of those activities and functions that absolutely have to be done by that group functioning as a group. Next, a list can be made of things that are not being done but need to be. A third list can be made of those things that the group is now doing that could be done, even if not so well, by either the same people wearing their functional or other hats, or by other people. Experience has shown that the second two lists tend to balance each other and tend to represent somewhere around 25-30 percent of the total work of such a group. Based on this analysis a replan of work can emerge. It can have significant effects on both attitudes and behavior. The output of such an activity by a group at the top means that work gets reallocated to the next level, and thus a domino effect is set in motion which can result in significant change.

Another illustration concerns an organization-wide change effort to improve both the way work is done and the management of the work. The total staff of this very large organization was about forty thousand people. During a six month period, the total organization met in their work teams with the task of developing the criteria against which that team wanted the performance of their work unit to be measured. They then located their current performance against those criteria and projected their performance at a date about six months in the future against the same criteria. These criteria and projections were checked with senior management committees in each subsystem. If approved they became the work plan and basis for performance appraisal for that group.

With this one intervention the top management distributed the responsibility for managing the work to the people who were doing the work throughout the organization. The results of this program were a significant increase in productivity, significant cost reductions, and a significant change in attitudes and feelings of ownership among large numbers of employees, many of whom were previously quite dissatisfied with the state of things. Given this participative mode, it is most unlikely that any future management could successfully return to overcentralized control. Much latent energy was released and continues to be used by people all over the organization who feel responsible and appreciated for *their* management of *their* work.

Change in the Reward System

One significant organization problem concerns making the reward system consistent with the work. How often we see organizations in which someone in a staff department spends 90 percent of his time in assisting some line department; yet for his annual review his performance is evaluated solely by the head of his staff department, probably on 10 percent of his work. One result of this is that any smart person behaves in ways that please the individual who most influences his career and other rewards rather than those with whom he is working. Inappropriate reward systems do much to sabotage effective work as well as organization health.

An example of an intervention in this area follows. The vice president of one of the major groups in a very large company was concerned about the lack of motivation by his division general managers toward working with him on planning for the future of the business as a whole. He was equally concerned that the managers were not fully developing their own subordinates. In his opinion, this was blocking the managers' promotions. The vice president had spoken of these concerns many times. His staff had agreed that it was important to change, but their behavior was heavily directed toward maintaining the old priorities: meeting short-term profit goals. This group existed in an organization where the reward system was very clear. The chief executives in any sub-enterprise were accountable for their short-term profits. This was their most important assignment. Division managers knew that if they did not participate actively in future business planning, or if they did not invest energy in the development of subordinates, they would incur the group vice president's displeasure. They also knew, however, that if they did not meet their short-term profit objectives, they probably would not be around. The company had an executive incentive plan in which considerable amounts of bonus money were available to people in the upper ranks for good performance. In trying to find a method for changing his division managers' priorities, the group vice president looked, with consultant help, at the reward system. As a result of this he called his colleagues together and told them, "I thought you'd like to know that in determining your bonus at the next review, I will be using the following formula. You are still 100 percent accountable for your short-term profit goals, but that represents 60 percent of the bonus. Another 25 percent will be my evaluation of your performance as members of this top management planning team. The other 15 percent will be your discernible efforts toward the development of your subordinates." Executive behavior changed dramatically. The reality of the reward system and the desired state were now consistent.

We have examined briefly several types of organization phenomena which need large system oriented interventions. We will now look at initial interventions and examine some of the choices facing the intervenor.

EARLY INTERVENTIONS

There are a number of choices about where to intervene. Several are listed here with the objective of creating a map of possibilities. The list includes:

1. The top team or the top of a system.
2. A pilot project which can have a linkage to the larger system.
3. Ready subsystems: those whose leaders and members are known to be ready for a change.
4. Hurting systems. This is one class of ready system where the environment has caused some acute discomfort in a generally unready system.
5. The rewards system.
6. Experiments: a series of experiments on new ways of organizing or new ways of handling communications.
7. Educational interventions: training programs, outside courses, etc.
8. An organization-wide confrontation meeting, bringing together a variety of parts of the organization, to examine the state of affairs and to make first step plans for improvement.[4]
9. The creation of a critical mass.

The last concept requires some elaboration. It is most difficult for a stable organization to change itself, that is, for the regular structures of the organization to be used for change. Temporary systems are frequently created to accomplish this. As an example, in one very large system, a country, there were a number of agencies involved in training and development for organization leaders. The government provided grants to the agencies for training activities. These grants also provided funds to support the agency staffs for other purposes. Because of this condition each agency was developing programs for the same small clientele. Each agency kept innovations secret from its competitors.

In an attempt to move this competitive state toward a more collaborative one, a small group of people developed a "nonorganization" called the Association for Commercial and Industrial Education. It was a luncheon club. Its rules were the opposite of an ordinary organization's. It could make no group decisions, it distributed no minutes, no one was allowed to take anyone else to lunch, there were no dues, and there were no officers or hierarchy.

In this context it was possible for individuals from the various competing agencies to sit down and talk together about matters of mutual interest. After a couple of years it even became possible to develop a

[4]For one view of this, see Beckhard [2].

national organization development training project in the form of a four week course which was attended by top line managers and personnel people from all the major economic and social institutions in the country. Only this nonorganization could sponsor such a program. From this program a great many other linkages were developed. Today there is an entire professional association of collaborating change agents with bases in a variety of institutions, but with the capacity to collaborate around larger national problems.

MAINTAINING CHANGE

To maintain change in a large system it is necessary to have conscious procedures and commitment. Organization change will not be maintained simply because there has been early success. There are a number of interventions which are possible, and many are necessary if a change is to be maintained. Many organizations are living with the effects of successful short-term change results which have not been maintained.

Perhaps the most important single requirement for continued change is a continued feedback and information system that lets people in the organization know the system status in relation to the desired states. Some feedback systems that are used fairly frequently are:

1. Periodic team meetings to review a team's functioning and what its next goal priorities should be.

2. Organization sensing meetings in which the top of an organization meets, on a systematic planned basis, with a sample of employees from a variety of different organizational centers in order to keep apprised of the state of the system.

3. Periodic meetings between interdependent units of an organization.

4. Renewal conferences. For example, one company has an annual five-year planning meeting with its top management. Three weeks prior to that meeting the same management group and their wives go to a retreat for two or three days to take a look at themselves, their personal and company priorities, the new forces in the environment, what they need to keep in mind in their upcoming planning, and what has happened in the way they work and in their relationships that needs review before the planning meeting.

5. Performance review on a systematic, goal-directed basis.

6. Periodic visits from outside consultants to keep the organization leaders thinking about the organization's renewal.

There are other possible techniques but this list includes the most commonly used methods of maintaining a change effort in a complex organization.

SUMMARY

In order to help organizations improve their operational effectiveness and system health, we have examined:

1. A model for determining early organization interventions.
2. Some choices of change strategies.
3. Some choices of early interventions.
4. Some choices of strategies for maintaining change.

The focus of this article has been on what the third party, facilitator, consultant, etc., can do as either a consultant, expert, trainer, or coach in helping organization leaders diagnose their own system and plan strategies for development toward a better state. This focus includes process intervention but is not exclusively that. It also includes the skills of system diagnosis, of determining change strategies, of understanding the relationship of organizations to external environments, and of understanding such organizational processes as power, reward systems, organizational decision making, information systems, structural designs and planning.

It is the author's experience that the demand for assistance in organizational interventions and large system organization change is increasing at a very fast rate, certainly faster than the growth of resources to meet the demand. As the world shrinks, as there are more multinational organizations, as the interfaces between government and the private sector and the social sector become more blurred and more overlapping, large system interventions and the technology and skill available to facilitate these will be in increasingly greater demand.

REFERENCES

[1] Beckhard, R. *Organization Development: Strategies and Models*. Reading, Mass.: Addison-Wesley, 1969.

[2] Beckhard, R. "The Confrontation Meeting." *Harvard Business Review*, March-April 1967.

[3] Krone, C. "Open Systems Redesign." In *Theory and Method in Organization Development: An Evolutionary Process*, edited by John Adams. Rosslyn, Virginia: NTL Institute, 1974.

[4] Lawrence, P. R., and Lorsch, J. W. *Organization and Environment: Managing Differentiation and Integration*. Boston: Harvard Business School, Division of Research, 1967.

The Coming Crisis for Production Management: Technology and Organization

L. E. Davis

INTRODUCTION

Events cast their shadows before them. Already, we can discern changes in our environment more than sufficient to show that Western industrial society is in transition from one historical era to another. It is the purpose of this paper to indicate that the environmental characteristics of the post-industrial era will lead to crisis and massive dislocation unless adaptation occurs. The anticipated consequences will be greatest, at first, for the production industries, because they stand at the confluence of changes involving technology, social values, the economic environment, organizational design, job design and the practices of management.

Managers, as rational leaders, will seek to avoid these consequences by altering the forms of institutional regulation and control. It is a secondary purpose of this paper to describe some ways in which managers are already beginning this process. Specifically, examples will be given from the research results and organizational experiments of an international coalition of English, American and Norwegian researchers whose reports are referred to throughout this paper.

Presented at the International Conference on Production Research, Birmingham, England, April 1970.

Reprinted from L. E. Davis, "The coming crisis for production management: technology and organization." *International Journal of Production Research*, 1971, 9(1), 65-82. Used with permission of the publisher, Taylor & Francis Ltd., London.

Before examining the modifications in organization and management that will be required to respond successfully to the evolving environmental changes, a brief review of the history and present state of production organization will be undertaken. How did jobs, organizations and managerial practices get to be what they are today? The present state can then be compared with the requirements flowing from evolving environmental changes.

CONVENTIONAL INDUSTRIAL ORGANIZATION

Any history of production organization, its jobs and management, is a reflection of a culture and the developed states of that culture's technology. Culture includes the belief systems and values held about individuals as members of a society—particularly, in this context, as working members. It is culture that sets bounds on the technology that may be applied to achieve desired goals and on the ways in which an organization is permitted to use that technology.

In discussing production organization, the Protestant Ethic is, of course, central. Work is the proper measure of a man's success; indeed, it provides him with his *raison d'etre*. Of nearly equal importance in the development of the production system has been the view of man as an operating unit, an element in an operating system (see Boguslaw 1965). As operating units, men can be adjusted, modified, changed by training, incentives, etc., for the good of society or to suit the needs of an organization. Within certain very broad limits, and as long as economic goals are satisfied, the individual and his needs are of secondary concern and, at best, simply a constraint.

A third crucial belief concerns the reliability and responsibility of individuals. Industrial society's view of individuals as unreliable fostered the development of the concept of men as spare parts. Individuals are assigned work in such a way that they can be treated as interchangeable. Each man is given only a single thing to do; many men are available ('in inventory', as it were) to do that thing if the first man fails. The human operating units are given narrow tasks of responsibilities and are seen as having narrow capabilities and small utility to the organization. Most current industrial training schemes are based on this spare-parts concept.

In fact, so much of industry is organized along these lines that any other approach becomes difficult to visualize, but consider for a moment those segments of society in which individuals organize their own worlds—notably in the professions, the arts and a tiny remnant of the crafts. Here, the governing concept is that redundancy resides not in the number of units (each capable of a single response), but in the number of responses that can be made by each unit. Where this principle rules (we may call it the 'spare-responses' concept), it postulates the necessity for each worker to possess a large repertoire of skills and capabilities so that he can adapt to the requirements of his work situation.

Three other beliefs are relevant here as well. The first is that labour is a commodity to be sold by individuals and purchased by organizations. In this view, individuals need not be considered as members of the organization, who may have changing contributions to make over time and other needs to satisfy. Common agreement on this belief may well explain why organized labour has not been concerned with the quality of work life. A second belief concerns materialism in its narrow sense, under which the ends of achieving higher economic standards or material benefits justify the means by which they are achieved. Lastly, there is the belief in the job as a disjoint increment—an isolated event in the life of the individual. This non-careerism syndrome is a central feature of life in industrial societies.

Technologies developed during the industrial era have had crucial effects on the organization of work. Technology may be defined as the complex of techniques, associated machinery and tools used to transform materials and information in a predicted way to achieve specified outcomes. The production technologies of the industrial era began with a number of developments in England about 150 years ago: (1) men and animals were replaced as the essential power sources for carrying out work; (2) the nature of the new power source required that groups of men be brought together around it—giving rise to the factory system; (3) some elements of men as workers (i.e. some of their skills) were displaced into mechanical tools and devices; (4) men and machines had to be coordinated; (5) the foregoing list of changes provided new opportunities to organize or 'rationalize' the ways in which men worked.

The central property of industrial technologies is that they are deterministic. What is to be done, how it is to be done, and when it is to be done are all specifiable. Cause and effect relationships are known, and all actions can be prescribed to obtain the desired results. Among the driving forces in the continuing development of production technology has been the attempt to make more and more processes completely specifiable, or, as will be indicated elsewhere, programmable.

Organizations evolving out of the design processes of allocating tasks to men and machines and developing a guiding and regulating superstructure reflected both the deterministic technology and the values and beliefs of Western industrial society. A new kind of specialization of labour was introduced in which jobs were deliberately fractionated so that unskilled people could do them. In 1835, when the practice of job fractionation had existed for some 20 years, Charles Babbage reported this social innovation in his book, *On the Economy of Machinery and Manufacture*, which is still relevant today. In about 1890, F. W. Taylor rediscovered what Babbage had done 100 years earlier and created an approach called 'scientific management' which is the present basis of industrial organization throughout the Western world.

Scientific management, as developed by Taylor, can be called the 'machine theory of organization'. It was characterized by the following elements:

1. The unit comprising the man and his job is the essential building block of an organization; if the manager gets this 'right' (in some particular but unspecified way), then the organization will be correctly defined.
2. Man is an extension of the machine, useful only for doing things that the machine cannot do.
3. The men and their jobs—the individual building blocks of organizations—are to be glued together by supervisors who will absorb the uncertainties and variabilities that arise in the work situation. Further, these supervisors need supervisors, and so on, *ad infinitum*, leading to the present many-layered hierarchy. In bureaucratic organizations, the latter notion ultimately leads to situations in which the job title of 'manager' is granted solely on the grounds that someone supervises a specified number of people.
4. The organization is free to use any available social mechanisms to enforce compliance and ensure its own stability, such as tighter task definitions, pacing of work, incentives, external supervision, etc.
5. Job fractionation is a way of reducing the costs of carrying on the work by reducing the skill contribution of the individual who performs it. Man is simply an extension of the machine, and by machine-theory logic, the more the machine is simplified (whether it is a living or non-living part) the more costs are lowered.
6. Attention to the primacy of technological requirements provides an optimum outcome; conversely, satisfying the requirements of interrelated social systems increases costs.
7. Technology, as science, is value-free; concomitantly, the design and planning of production technologies are norm- and value-free.

Characteristically, present production organizations are based to a large degree on the machine theory of organization in which interdependence between tasks and between individuals is controlled by special managerial arrangements, systems of payment, etc. Such organizations have large superstructures designed to coordinate the elements in which work is done, join them together, counteract variances arising both within the elements and within the socio-organizational links created by its members, and adjust the system to changes in input or output requirements. In such organizations, thinking, planning, coordinating and controlling are functions exercised within the superstructure; transformation tasks, most of which are programmable, are performed at the worker levels. The consequences of such organizational arrangements are too well known to bear repeating. Characteristically, management is reinforced in its beliefs that workers are unreliable, interested only in external rewards, and regard their work as a burden to be set aside at the first possible opportunity. Largely, this is a self-fulfilling prophecy. What saves the day is that the organizational system can be maintained—rickety though it is—as long as the technology remains deterministic and social expectations for a humane quality of work life are not too widespread.

THE POST-INDUSTRIAL CHALLENGE

Changes in Society

In recent years, changes in Western societal environments have been reflective of a rising level of expectations concerning material, social and personal needs. The seeming ease with which new (automated) technology satisfies material needs, coupled with the provision of subsistence-level support for its citizens by society, has stimulated a growing concern on the part of the individual over his relationship to work, its meaningfulness and its value—i.e. a concern for the quality of work life (see Davis 1970 a). In the U.S., questioning of the relationship between work and satisfaction of material needs is widespread through the ranks of university students, industrial workers and minority unemployed. The viability of the belief, already described, that individuals may be used to satisfy the economic goals of organizations is being seriously questioned. It appears that people may no longer let themselves be used; they wish to see some relationship between their own work and the social life around them, and they wish some desirable future for themselves in their continuing relationship with organizations. No longer will workers patiently endure dehumanized work roles in order to achieve increased material rewards.

Among university students these expectations are leading to refusals to accept jobs with major corporations, in favour of more 'socially oriented' institutions—an unfortunate loss of talented people. Even the unemployed are refusing to accept dead-end demeaning jobs (see Doeringer 1969), appearing to be as selective about accepting jobs as are the employed about changing jobs. There appear to be means, partly provided by society, for subsisting in minority ghettos without entering the industrial world. For industrial workers there is a revival of concern with the once-buried question of alienation from work, job satisfaction, personal freedom and initiative, and the dignity of the individual in the work place. Although on the surface the expressed concern is over the effects of automation on job availability and greater sharing in wealth produced, restlessness in unions, their failure to grow in the nonindustrial sectors and the frequent overthrow of union leaders are all indicators, in the U.S., of a changing field that stems from the increasingly tenuous relationship between work and satisfaction of material needs.

Another factor impelling social change is the continuously rising level of education that Western countries provide, which is changing the attitudes, the aspirations, and the expectations of major segments of society (see Bell 1967). Future trends are already visible in California, where almost 50% of young people of college and university age are in school and where one-third of all the scientists and engineers in the U.S. are employed.

Changes in Technology

One of the forces driving the transition into the post-industrial era is the growing application of automated, computer-aided production systems. This development is bringing about crucial changes in the relationship between technology and the social organization of production—changes of such magnitude that the displacement of men and skills by computers is reduced to the status of a relatively minor effect.

The most striking characteristic of sophisticated, automated technology is that it absorbs routine activities into the machines, creating a new relationship between the technology and its embedded social system; the humans in automated systems are interdependent components required to respond to *stochastic*, not deterministic, conditions—i.e. they operate in an environment whose 'important events' are randomly occurring and unpredictable. Sophisticated skills must be maintained, though they may be called into use only very occasionally. This technological shift disturbs long-established boundaries between jobs and skills and between operations and maintenance. It has also contributed to a shift in the working population from providing goods to providing personal and societal services. As may be expected, there is a shift from blue-collar to white-collar work in clerical, technical and service jobs. At all levels of society, individuals find that they must change their careers or jobs over time.

Still further, the new technology requires a high degree of commitment and autonomy on the part of workers in the automated production processes (see Davis 1970 b). The required degree of autonomy is likely to be in serious conflict with the assumptions and values held within the bureaucratic technostructure (see Galbraith 1967).

Another feature is that there are in effect two intertwined technologies. The primary technology contains the transformations needed to produce the desired output. It is machine- and capital-intensive. The secondary technology contains the support and service activities, such as loading and unloading materials, tools, etc. It is labour-intensive and its variances are capable of stopping or reducing throughput, but enhancing the secondary technology will not enhance the primary technology and its throughputs.

Although it poses new problems, highly sophisticated technology possesses an unrecognized flexibility in relation to social systems. There exists an extensive array of configurations of the technology that can be designed to suit the social systems desired, within limits. This property disaffirms the notion of the 'technological imperative' widely held by both engineers and social scientists. It places the burden on managers, hopefully aided by social scientists, to elucidate the characteristics of their particular social system suitable to the evolving post-industrial era.

Changes in Economic Organization

Developments in technology are interrelated with changes in economic organization. The scale of economic units is growing, stimulated by the developments of sophisticated production technology and organized knowledge leading to new products. In turn this is leading to new arrangements in the market, stimulating the development of higher-order interactions.

The organized use of knowledge brings about constant product innovation and for firms in electronics, aerospace, computers, information processing, etc., a new phenomenon in market relationship appears. Such firms are continually in the process of redefining their products and their futures—an exercise that reflects back on their internal organization structures and on the response flexibility of their members. Within these companies, there is an observable shift to high-talent personnel and to the development of strategies of distinctive competence, stores of experience, and built-in redundancy of response capabilities.

The Consequences of These Changes

A pervasive feature of the post-industrial environment is that it is taking on the quality of a turbulent field (see Emery and Trist 1965). Turbulence arises from increased complexity and from the size of the total environment. It is compounded by increased interdependence of the environment's parts and the unpredictable connections arising between them as a result of accelerating but uneven change. The area of relevant uncertainty for individuals and organizations increases and tests the limits of human adaptability; earlier forms of adaptation, developed in response to a simpler environment, appear to suffice no longer. The turbulent environment requires that boundaries of organizations be extended into their technological, social and economic environments. The organization needs to identify the causal characteristics of the environments so that it can develop response strategies. The production organization, in particular, must provide a structure, a style of management and jobs so designed that adaptation can take place without massive dislocation.

THE POST-INDUSTRIAL OPPORTUNITY

Although the presence of the features outlined in the previous section indicates that we are already well launched into the post-industrial era, Trist (1968) finds that we suffer from a cultural lag—the absence of a culture congruent with the identifiable needs of post-industrialism. Furthermore, in the turbulent environmental texture of the post-industrial era, the individual organization, city, state, or even nation—acting alone—may be unable to meet the demands of increasing levels of complexity. Resources will have to be pooled; there will need to be more sharing, more trust and more cooperation.

Seldom does society have a second chance to redress deep-seated errors in social organization and members' roles; however, the opportunity may now be at hand to overcome alienation and provide humanly meaningful work in socio-technical institutions (see Fromm 1968 and Emery 1967). The development, over a period of nearly 20 years, of a body of theory (see Emery 1969) concerned with the analysis and design of interacting technological and social systems has furthered the examination of questions of organization and job design in complex environments, too long considered to be exclusively an art form. The diffusion of knowledge about applications of these theories is itself changing the environment of other organizations. The concepts were first developed in Britain (see Emery and Trist 1960) and followed by developments in the United States and recently in Norway, Canada and

Sweden. They are far from having come into common practice. Their most comprehensive application is taking place in Norway, on a national scale, as a basis for developing organizational and job design strategies suitable to a democratic society.

Briefly, socio-technical systems theory rests on two essential premises. The first is that in any purposive organization in which men are required to perform the organization's activities, there is a joint system operating, a *socio-technical* system. When work is to be done, and when human beings are required actors in the performance of this work, then the desired output is achieved through the actions of a social system as well as a technological system. Further, these systems so interlock that the achievement of the output becomes a function of the appropriate joint operation of both systems. The operative work is 'joint', for it is here that the socio-technical idea departs from more widely held views—those in which the social system is thought to be completely dependent on the technical system. The concept of joint optimization is proposed, which states that it is impossible to optimize for overall performance without seeking to optimize jointly the correlative independent social and technological systems.

The second premise is that every socio-technical system is embedded in an environment—an environment that is influenced by a culture and its values, an environment that is influenced by a set of generally acceptable practices, an environment that permits certain roles for the organisms in it. To understand a work system or an organization, one must understand the environmental forces that are operating on it. Without this understanding, it is impossible to develop an effective job or organization. This emphasis on environmental forces suggests, correctly, that the socio-technical systems idea falls within the larger body of 'open system' theories. What does this mean? Simply, that there is a constant interchange between what goes on in a work system or an organization and what goes on in the environment; the boundaries between the environment and the system are highly permeable, and what goes on outside affects what goes on inside. When something occurs in the general society, it will inevitably affect what occurs in organizations. There may be a period of cultural lag, but sooner or later, the societal tremor will register on the organizational seismographs.

Significantly, socio-technical systems theory provides a basis for analysis and design overcoming the greatest inhibition to development of organization and job strategies in a growing turbulent environment. It breaks through the long-existing tight compartments between the worlds of those who plan, study and manage social systems and those who do so for technological systems. At once it makes nonsensical the existing positions of psychologists and sociologists that in purposive organizations the technology is unalterable and must be accepted as a given requirement. Most frequently, therefore, only variables and relationships not influenced by technology are examined and altered. Without inclusion of technology, which considerably determines what work is about and what demands exist for the individual and organization, not only are peripheral relations examined but they tend to become disproportionately magnified, making interpretation and use of findings difficult, if not impossible. Similarly, it makes nonsensical the 'technological imperative' position of engineers, economists and managers who consider psychological and social requirements as constraints and at best as boundary conditions of technological systems. That a substantial part of technological system design includes social system design is neither understood nor appreciated. Frightful assumptions, supported by societal values, are made about men and groups and become built into machines and processes as requirements.

Socio-technical systems analysis provides a basis for determining appropriate boundaries of systems containing men, machines, materials and information. It considers the operation of such systems within the framework of an environment that is made an overt and specific object of the socio-technical study. It concerns itself with spontaneous reorganization or adaptation, with control of system variance, with growth, self-regulation, etc. These are aspects of system study that will become increasingly important as organizations in the post-industrial era are required to develop strategies that focus on adaptability and commitment. For these reasons, socio-technical systems analysis is felt to offer one of the best current approaches to meeting the post-industrial challenge.

The final section of this paper presents some selective aspects of sociotechnical theory and application in greater detail. Wherever possible, actual field studies using the socio-technical approach are cited to support and illustrate the discussion.

RESULTS OF ORGANIZATIONAL AND JOB DESIGN RESEARCH

A number of developments, including on-site organizational experiments, lend strong support to the prospects of successfully developing suitable strategies of organization for the post-industrial era. In general, successful outcomes as measured by various objective criteria depended on finding an accommodation between the demands of the organization and the technology on the one hand, and the needs and desires of people on the other, so that the needs of both were provided for. A summary report of U.S. and English empirical studies appeared in Davis (1966).

The studies sought to find conditions in organization structure and job contents leading to cooperation, commitment, learning and growth, ability to change, and improved performance. The findings can be summarized under four categories of requirements: responsible autonomy, adaptability, variety and participation. When these factors were present, they led to learnings and behaviours that seemed to provide the sought-for organization and job response qualities. These studies lend support to the general model of responsible autonomous job and group behaviour as a key facet in socio-technological relationships in production organizations.

By autonomy is meant that the content, structure and organization of jobs are such that individuals or groups performing those jobs can plan, regulate and control their own worlds. Autonomy implies a number of things, among which may be the need for multiple skills within the individual or within a group organized so it can share an array of skills. Also implied are self-regulation and self-organization, which are radical notions in conventional industrial organization. Under the principle of self-regulation, only the critical interventions, desired outcomes and organizational maintenance requirements need to be specified by those managing, leaving the remainder to those doing. Specifically, situations were provided in which individuals or groups accepted responsibility for the cycle of activities required to complete the product or service. They established the rate, quantity and quality of output. They organized the content and structure of their jobs, evaluated their own performance, participated in setting goals and adjusted conditions in response to work-system variability.

The results obtained indicated that when the attributes and characteristics of jobs were such that the individual or group became largely autonomous in the working situation, then meaningfulness, satisfaction and learning increased significantly, as did wide knowledge of process, identification with product, commitment to desired action and responsibility for outcomes. These supported the development of a job structure that permitted social interaction among job-holders and communication with peers and supervisors, particularly when the maintenance of continuity of operation was required. Simultaneously, high performance in quantity and quality of product or service outcomes was achieved. This has been demonstrated in such widely-different settings as the mining of coal (reported by Trist and Emery 1963), the maintenance of a chemical refinery, and the manufacture of aircraft instruments (reported by Davis and Werling 1960 and Davis and Valfer 1966).

The second requirements category, which has mainly been the province of psychologists, is concerned with 'adaptation'. The contents of the job have to be such that the individual can learn from what is going on around him, can grow, can develop, can adjust. Slighted, but not overlooked, is the psychological concept of self-actualization or personal growth, which appears to be central to the development of motivation and commitment through satisfaction of higher order intrinsic needs of individuals. The most potent way of satisfying intrinsic needs may well be through job design (see Lawler 1969). Too often jobs in conventional industrial organizations have simply required people to adapt to restricted, fractionated activities, overlooking their enormous capacity to learn and adapt to complexity. (Such jobs also tend to ignore the organization's need for its workers to adapt.) In sophisticated technological settings, the very role of the individual is dependent on his adaptability and his commitment. With nobody around at the specific instant to tell him what to do, he must respond to the situation and act as needed. The job is also a setting in which psychic and social growth of the individual should take place. Blocked growth leads to distortions that have costs for the individual, the organization and society.

Where the socio-technical system was so designed that the necessary adaptive behaviour was facilitated, positive results in economic performance and in satisfactions occurred at all levels in the organization, as demonstrated in studies in oil refineries, automated chemical plants, pulp and paper plants (see Thorsrud and Emery 1969), and aircraft instrument plants (see Davis 1966).

The third category is concerned with variety. Man, surely, has always known it, but only lately has it been demonstrated that part of what a living organism requires to function effectively is a variety of experiences. If people are to be alert and responsive to their working environments, they need variety in the work situation. Routine, repetitious tasks tend to extinguish the individual. He is there physically, but not in any other way; he has disappeared from the scene. Psychologists have also studied this phenomenon in various 'deprived environments'. Adult humans confined to 'stimulus-free' environments begin to hallucinate. Workers may respond to the deprived work situation in much the same way—by disappearing (getting them back is another issue). Variety in industrial work has been the subject of study and controversy for 50 years. Recently, considerable attention has focused on the benefits to the individual and the organization of enlarging jobs to add variety (see Herzberg 1966 and Davis 1957).

There is another aspect of the need for variety that is less well-recognized in the industrial setting today, but that will become increasingly important in the emergent technological environment. The cyberneticist, Ashby (1960) has described this aspect of variety as a general criterion for intelligent behaviour of any kind. To Ashby, adequate adaptation is only possible if an organism already has a stored set of responses of the requisite variety. This implies that in the work situation, where unexpected things will happen, the task content of a job and the training for that job should match this potential variability.

The last category concerns participation of the individual in the decisions affecting his work. Participation in development of job content and organizational relations, as well as in planning of changes, was fundamental to the outcomes achieved by the studies in Norway (see Thorsrud and Emery 1969) and in the aircraft instrument industry (see Davis, 1960 and Davis 1966). Participation plays a role in learning and growth and permits those affected by changes in their roles and environments to develop assessments of the effects. An extensive literature on the process and dynamics of change (see Bennis 1966) supports the findings of the field studies.

In a pioneering study, Lawrence and Lorsch (1967) examined the effects of uncertainty in technology and markets on the structure, relationship and performance of organizations. They found that where uncertainty is high, influence is high, i.e. if the situation becomes increasingly unpredictable, decision-making is forced down into the organization where the requisite expertise for daily decisions resides. Under environments of uncertainty, influence and authority are more evenly distributed; organizations become 'polyarchic'. Under environments of certainty or stability, organizations tend to be relatively less democratic, with influence, authority and responsibility centralized. These findings were derived from studies of firms in contrasting certain and uncertain environments.

Another category, which goes beyond the four and was implicit in them, concerns the total system of work. In the field studies, if tasks and activities within jobs fell into meaningful patterns, reflecting the interdependence between the individual job and the larger production system, then enhanced performance satisfaction and learning took place. In socio-technical terms, this interdependence is most closely associated with the points at which variance is introduced from one production process into another. When necessary skills, tasks and information were incorporated into the individual or group jobs, then adjustments could be made to handle error and exceptions within the affected subsystem; failing that, the variances were exported to other interconnecting systems. (In 'deterministic' systems, the layer on layers of supervisors, buttressed by inspectors, utility men, and repairmen, etc., absorb the variances exported from the work-place.)

These organizational experiments indicate that individuals and organizations can change and adapt to turbulent environments. Nonetheless, in moving into the post-industrial era, considerable learning is still needed about building into the organizational milieu the capability for continuing change. A number of studies have indicated that, if spontaneous and innovative behaviours are to result, conditions will have to be developed to bring about internalization of organizational goals (see Katz and Kahn 1966). Such internalization exists at the upper levels of organizations, but (except in the Norwegian experiments) is found in the lower levels only in voluntary organizations.

CONCLUSION

In the post-industrial era, current organization structures will become increasingly dysfunctional. If strategies of survival are to be developed, advanced societies, particularly the managers of their industrial and business organizations, will have to accept the obligation to examine existing assumptions and face the value issues regarding men and technology raised by the evolving environments. Existing jobs and organizations will have to undergo reorganization to meet the requirements for a continuing high rate of change, new technologies and changing aspirations and expectations. These undertakings will be wrenching for institutions and individuals. Providing prescriptions would be presumptuous, but some organizations, joined by sociotechnical researcher-consultants, seem to be well into the process.

REFERENCES

ASHBY, W. R., 1960, *Design for a Brain* (New York: Wiley).

BELL, D., 1967, Notes on the post-industrial society (I and II), *Publ. Interest*, Nos. 6 and 7.

BENNIS, W. G., 1966, *Changing Organisations* (New York: McGraw-Hill).

BOGUSLAW, R., 1965, *The New Utopians* (New York: Prentice-Hall), Chap. 5.

BURNS, T., and STALKER, G., 1961, *Management of Innovation* (London: Tavistock Publications).

CROOME, H., 1960, *Human Problems of Innovation (London: H.M.S.O.).*

DAVIS, L. E., 1957, Toward a theory of job design, *J. ind. Engng*, **8**, 305; 1962, The effects of automation on job design, *Ind. Relat.*, **2**, 53; 1966, The design of jobs, *Ibid.*, **6**, 21; 1970a, Restructuring jobs for social goals, *Manpower*, **2**, 2; 1970b, Job satisfaction—a sociotechnical view, *Ind. Relat.*, **10**.

DAVIS, L. E., and VALFER, E. S., 1966, Studies in supervisory job design, *Hum. Relat.*, **17**, 339.

DAVIS, L. E., and WERLING, R., 1960, Job design factors, *Occup. Psychol.*, **28**, 109.

DOERINGER, P. B., 1969, Ghetto labor markets and manpower, *Mon. lab. Rev.*, **55**.

EMERY, F. E., 1967, The next thirty years: concepts, methods and anticipations, *Hum. Relat.*, **20**, 199; 1969, *Systems Thinking* (London: Penguin Books).

EMERY, F. E., and THORSRUD, E., 1969, *Form and Content in Industrial Democracy* (London: Tavistock Publications).

EMERY, F. E., and TRIST, E. L., 1960, *Management Sciences, Models and Techniques*, Vol. II (London: Pergamon), p. 83; 1965, The causal texture of organisational environments, *Hum. Relat.*, **18**, 21.

FROMM, E., 1968, *The Revolution of Hope: Toward a Humanised Technology* (New York: Harper & Row), Chap. 5.

GALBRAITH, J. K., 1967, *The New Industrial State* (Boston: Houghton-Mifflin).

HERZBERG, F., 1966, *Work and the Nature of Man* (New York: World).

KATZ, D., and KAHN, R. L., 1966, *Social Psychology of Organisations* (New York: Wiley), p. 345.

LAWLER, E. E., 1969, Job design and employee motivation, *Personn. Psychol.*, **22**, 426.

LAWRENCE, P. R., and LORSCH, J. H., 1967, *Organisation and Environment* (Cambridge: Harvard University Press).

PERROW, C., 1967, A framework for comparative analysis of organisation, *Am. sociol. Rev.*, **32**, 194.

THORSRUD, E., and EMERY, F., 1969, *Moten NY Bedriftsorganisasjon* (Oslo: Tanum Press), Chap. 6.

TRIST, E. L., 1968, *Urban North America, The Challenge of the Next Thirty Years–A Social Psychological Viewpoint* (Town Planning Institute of Canada).

TRIST, E. L., and EMERY, F. E., 1963, *Organisational Choice* (London: Tavistock Publications).

Sociotechnical Systems: An Intervention Strategy

T. G. Cummings

Since its beginnings in the early 1950s, sociotechnical systems theory has evolved as a primary means of confronting and dealing with problems of alienation of workers, apathy, and inferior quality of work. Starting with the innovative experiments in the British coal mines (Trist & Bamforth, 1951) and continuing with job design research in the United States (Davis & Canter, 1955), the sociotechnical systems approach has gained increasing acceptance among researchers and practitioners of organizations, especially those in industry.

The purpose of this chapter is to present a strategy for sociotechnical system intervention. The strategy is an attempt to introduce principles of sociotechnical analysis and design into an organization context. A brief description of sociotechnical systems theory will be presented initially. This is followed by a detailed discussion of each stage of the intervention strategy, and finally, conclusions about its general utility for organizational change are presented.

SOCIOTECHNICAL SYSTEMS THEORY

Sociotechnical systems theory is concerned with those organizational settings in which human beings are required to perform tasks in order to produce desired results. Hospitals, schools, industrial factories, service

---

Reprinted from T. G. Cummings, "SOCIOTECHNICAL SYSTEMS: An Intervention Strategy." In W. Warner Burke, *Current Issues and Strategies in Organization Development*, pp. 187–213, © 1976. Used by permission of Human Sciences Press, 72 Fifth Avenue, New York, N.Y. 10011.

organizations, and the like are prime examples of such work systems. The term *sociotechnical system* implies two fundamental concepts. The first is the joint operation of two independent but correlative systems: a social system composed of human beings who are the required actors in the performance of work, and a technological system made up of the tools, techniques, and methods of doing that are employed in task accomplishment. The achievement of a desired outcome requires the joint operation of both systems. Specifically, it is proposed that "it is impossible to optimize for overall performance without seeking to optimize jointly the correlative but independent social and technological systems" (Davis & Trist, 1972, p. 3). The term *joint optimization* refers to those conditions in which the requirements of the technology and the social and psychological needs of the workers are jointly met.

The second concept is that a sociotechnical system continually interacts with an environment which both influences and is influenced by the work system. When viewed as an open system, a sociotechnical system exists and grows only to the extent that it maintains viable interchanges with its environment. Since these interchanges involve a two-way flow of information and certain forms of matter and energy between the system and its environment, one must understand the dynamics of this relationship in order to understand the functioning of a work system. An additional requirement for optimal performance is that a sociotechnical system must maintain viable relationships with a succession of suitable environments if it is to survive and grow.

In summary, sociotechnical systems theory provides a basis for analyzing and designing work systems so that the social and technological systems are jointly optimized and the system as a whole is "capable of integrating proactively with its physical, social and biological environment" (Clark & Krone, 1972, p. 284).

A SOCIOTECHNICAL INTERVENTION STRATEGY

The introduction of new organizational ideas always involves some form of implicit or explicit change strategy. Since most attempts to apply sociotechnical systems theory are carried out in the context of an existing organization, it is important to make as explicit as possible a framework for implementing this approach in such settings. In this respect, the change strategy offered here is aimed at those organizational members who desire to experiment with a sociotechnical approach in their organizations. The strategy is based on three essential assumptions. The first involves the use of an experiment in a small part of the organization as a means of introducing the approach to the organization. By experimenting with sociotechnical principles under relatively protected experimental conditions, it is possible to gain a great deal of understanding and to make informed decisions about the utility of the approach for other parts of the organization. An experimental strategy for introducing new innovations is quite similar to an agricultural model in which field experiments are carried out at various demonstration sites connected with agricultural experiment stations. New ideas are first tried out under experimental conditions; then those that have some merit are examined under normal working circumstances; and finally, the results are disseminated to interested farmers by county extension agents who serve as change agents for farming practices. An experimental strategy is especially relevant for sociotechnical change. It allows organizations to test various configurations of social and technological structures in a limited part of the organization, thereby reducing the amount of overall disruption; it enables organizations to examine overall effects of new designs, thereby increasing their understanding of work-system functioning; and it gives organizational members the chance to see how the approach works in a real-life situation before accepting or rejecting it.

The second assumption is concerned with the organizational climate that is required for successful sociotechnical change. Most forms of organizational change require a certain degree of openness, risk taking, and experimenting in interpersonal and group relations if change is to be an accepted part of organizational life. From top management down to the shop floor, organizational members who are open, experimental, risk taking, and high in trust are better able to confront and work through the myriad of social problems that arise when structural change occurs. Since these changes affect the power structure, the reward system, and the status and recognition network of organizations, it is crucial that organizational members be open to and capable of dealing with these issues. In short, the more a change program requires behavioral changes that are deviant from existing norms, are personally discomforting, and entail high degrees of change from traditional methods, the more it is necessary to focus on interpersonal and group dynamics (Argyris, 1972).

The final assumption involves organizational members who carry out change. Since sociotechnical experimentation often takes place at lower organizational levels, it is imperative that those for whom change is intended be actively engaged in the change process. This not only increases the likelihood that valid information is attainable, but also provides a realistic basis for redesign proposals and subsequent change. In this regard, sociotechnical change emerges from inside the work system; it evolves from workers who operate the technological processes of the organization. This implies a "bottom-up" method of change. Top management sanction and support is required for such efforts, however.

The sociotechnical intervention strategy, then, is based on these assumptions. It is also grounded on work performed by the Human Resources Center at the Tavistock Institute, London, England (Rice, 1968; Trist et al., 1963) and by the Work Research Institute, Oslo, Norway (Thorsund, 1966), in addition to current field experiments carried out by the author in conjunction with the Organizational Behavior Group at Case Western Reserve University. The strategy consists of eight sequential stages: (1) defining an experimental system; (2) sanctioning an experiment; (3) establishing an action group; (4) analyzing an experimental system; (5) generating hypotheses for redesign; (6) testing and evaluating hypotheses for redesign; (7) transferring to a normal operating system; and (8) disseminating the results.

Defining an Experimental System

Defining a work system that is appropriate for purposes of sociotechnical experimentation is an extremely complex issue. Ideally, the unit should have the following characteristics:

1. Social and technological components that are clearly differentiated from other organizational units
2. Input and output states that are clearly defined and easily measurable
3. High probability for success
4. High potential for diffusion of results
5. Members who are interested in experimentation

In determining if a work system is *clearly differentiated from other organizational units*, one is faced with the problem of bounding a system—deciding what to include in the system and what not to include. Since this is always an arbitrary decision, one could conclude that it is a rudimentary process and not worthy of much attention. After all, why not take an existing organizational system and experiment with it? If existing work systems were defined so that their social and technological components formed a relatively self-completing whole, this argument would have some merit, for it would be possible to clearly differentiate between those processes that are relatively autonomous and under a system's control and those processes that are relatively heteronomous and not under its control. In sum, one could draw the distinction between system and environment.

Unfortunately, many organizational entities are not defined in terms of social and technological wholes; instead they are often differentiated around part processes in which the components are as much related to external components as they are to each other. While such work systems are meant to serve organizational needs for supervision and control—X number of workers for X number of supervisors—they pose difficult problems for change efforts. One can readily imagine the problems that arise when change is attempted in a work setting in which significant components that effect change are not included within the system's boundary. The crux of the argument is that many work units do not meet a minimal definition of a system: a distribution of components which function to maintain a superordinate whole in time and space. And when the purpose for defining a system is to change it, it is imperative that the components which relate to each other for the maintenance of a whole be included in the system definition.

Now that the importance of bounding a work system to include social and technological components that form a relatively self-completing whole has been established, the question arises as to how to identify such systems. A number of useful criteria exist to help identify a relatively differentiated sociotechnical system—time, territory, technology, and sociological and psychological attributes. Time refers to the contemporaneous existence of components—do they share the same time frame? Territory refers to the physical proximity of parts—do they share the same physical space? Technology is concerned with components that operate or share similar transformation processes—do they have a common conversion process? Sociological attributes pertain to individuals and have to do with work roles, supervision, and demographic characteristics—do workers have similar jobs, similar supervison, similar ages, religion, socioeconomic status, education, race, and sex? Finally, psychological attributes pertain to persons and their psychological states, including perceptual realities—do workers perceive themselves as a work system? Since sociotechnical systems contain both social and technological elements, the criteria help to account for the characteristics of each. Time, territory, and technology apply to both types of components, while sociological and psychological attributes pertain to the individuals who operate the work unit.

The problem of where to place a work system's boundaries is far from an elementary issue. While the criteria can help to define a relatively differentiated whole, the result is ultimately arbitrary. With this in mind, one tentatively bounds a work system using all the information and ingenuity available while realizing that the definition is open to change in light of new information.

A second characteristic of an appropriate experimental system is *clearly defined and easily measurable input and output states*. If the input and output conditions are not able to be determined, it is not possible to ascertain the functioning of a work system. Unless one can demonstrate that inputs are converted into outputs by particular actions, the rationality of a work unit's functioning remains unknown. The economic question of whether outputs are obtained in the least costly manner cannot be considered until it is shown that an outcome can be obtained. To the extent that inputs and outputs are explicitly defined and measured, the task of determining the effectiveness and efficiency of a work system's functioning is facilitated.

In choosing an experimental site with a *high probability of success*, it is necessary to decide upon appropriate criteria for judgment. Then, given these criteria, one must appraise the likelihood that a work unit can demonstrate positive results on these dimensions. Since the criterion problem is discussed later in this paper, let us assume that we have solved this issue and now must determine probability of success.

One strategy is to choose a work unit that scores extremely low on the criteria and assume that it has nowhere to go but up. Based on the premise that low scores are a valid measure of the system's present condition—that they are not extreme measures which are likely to statistically regress upward toward the unit's true performance, thereby producing a statistical artifact instead of actual change—this strategy has both positive and negative consequences. On the positive side, low performing systems are probably most in need of change, and carrying out a successful experiment can demonstrate dramatic results. On the negative side, such units can be so beset with problems that they do not have enough productive slack and energy to engage in experimentation. Also, the assumption that there is nowhere to go but up might be quite fallacious. The converse strategy of choosing a high-performing system and assuming that it is a healthy unit with a good base for improvement also has certain merits and limitations. On the plus side, the members of such work systems are often secure enough to experiment with change; they often seek challenges that have the potential of ensuring or improving their successful image, and they usually produce enough productive slack to allow for experimentation and improvement. On the other hand, high-performing systems can be an experimental risk. Existing social relationships can be harmed; productivity can decrease; system members can resent the implication that they should improve; the work system could be at its upper limits of performance.

Regardless of the strategy employed in choosing a potentially successful experimental site, the actual choice will have both intended and unintended consequences. The reality of any experiment is the uncertainty of obtaining desired results. In this respect, there is no optimal strategy for guaranteeing outcomes. Instead, one can explicitly examine the possible merits and limitations of one site versus another and proceed accordingly.

Since one of the purposes of a sociotechnical experiment is to generate findings that are applicable in other work settings, the choice of an experimental unit should take into account its *potential for diffusion of results*. In scientific terms, the system should have high generalizability; while in terms of change strategy, it should be a leading edge for innovation. The question of generalizability can never totally be resolved. No two work settings are exactly alike with respect to factors that can effect an experiment. Instead of dropping the issue here, there are a number of considerations that can increase the chances that results are transferable to other settings. It is possible to match a variety of dimensions across settings. Technology, social characteristics of system members, and environmental variables are a few of the factors that can be controlled for in the experimental site. By choosing a unit that is similar in these dimensions to other systems to which the results might be applied, the probability of generalizing to these other settings is increased.

Judging the extent to which a work system is a leading edge for innovation is crucial for a strategy in which an initial experiment is a starting condition for wider systemic change. Current theory is limited to the diffusion of new products and innovations. Studies on the dissemination of agricultural and medical innovations are prime examples. A relevant conclusion of these studies is that the individuals who are most likely to adopt new practices are those who most actively seek new knowledge. And, if this knowledge is to have wider acceptance, these individuals must also be influential. Applying this to sociotechnical diffusion, the experimental unit must not only seek new ways of doing things, but must have a key position in the influence structure of the organization. Determining the amount of influence a work unit possesses requires knowledge of the influence network of an organization. This entails an examination of the formal structure of influence that is exemplified in an organization chart as well as of the informal network along which unofficial sources of change flow.

A final characteristic of an ideal experimental system is that the *members are interested in experimentation*. Since sociotechnical experimentation is done in collaboration with members from the experimental unit, it is imperative that workers and managers be given the opportunity to make their own choices. Much like sensitivity training, it is difficult to present an accurate portrayal of an experiment in advance. Instead, pertinent information concerning the change strategy and the reasons for considering a particular unit for experimentation are given. At best, a tentative agreement among workers and managers to actively engage in the analysis and potential redesign of their work system is reached. At periodic intervals during the experiment, the issue of interest is raised, thereby leaving the possibility of termination of the experiment open for further negotiation. Without active participation by members of the work system, sociotechnical experimentation is devoid of any real significance.

Sanctioning an Experiment

Once an experiment site is chosen, terms for carrying out the experiment must be specified and sanctioned by workers and management. The primary purpose for sanctioning an experiment is to provide necessary protection for the experimental system. Normal organizational demands that impinge on the work unit must be temporarily reduced or suspended if new designs are to be tested under experimental conditions: workers must be afforded wage and job security if they are to feel free to experiment without the normal defenses that arise from fear of losing money or a job; and contractual arrangements such as job classification, wage scales, and the like must be open to further negotiation if hypotheses for redesign are not to be severely limited. Sociotechnical experiments require the same if not more protection than the start-up of a new work system. The primary means of providing necessary protection is by official sanction from the highest organizational levels that directly effect the experimental system. Sanctioning requires the active concern of workers—including union officials, if this is the case—and managers—both line managers and managers of other organizational units that directly relate to the experimental unit. Since the major function of the sanctioning body is to provide a protective umbrella under which experimentation can take place, it is imperative that its membership include organizational members that have control over variables that can effect the work system.

Establishing an Action Group

Sociotechnical experiments are often carried out under the direction of a small group composed of the workers and first-line management of the experimental unit. The purpose of the action group is to perform the analysis and generate hypotheses for redesign. It may also supervise the testing and evaluation phases of the experiment. In either case, the action group serves as the primary link between the experimental system and the sanctioning body. In performing its function, the action group collects and analyzes data, proposes new designs, and tests and evaluates results. It also keeps the sanctioning body informed of experimental progress and serves as a sounding board for involved workers and managers.

Action groups are utilized when the size of the experimental system is too large for all members to directly supervise the experiment—when the size of the unit exceeds eight to ten workers. Since experimental activities are often perceived as complex and difficult to understand, help is sometimes needed in the form of an internal or external consultant. Although the extent of such aid will vary according to the needs of the experimenters, the basic process of sociotechnical experimentation can be learned and carried out by most workers and managers.

In order to expedite the experimental process, the action group should be limited in size from three to six members—first-level supervision and a few interested and committed workers. During the analysis and hypothesis generation phase, some if not all members of the action group should spend full time carrying out experimental activities. This not only provides the necessary impetus for starting an experiment, but also serves as a positive sign of organizational support. Since the action group represents the experimental system, all efforts should be made to facilitate communication and influence between the action group, the sanctioning body, and the members of the experimental system.

Analyzing an Experimental System

The analysis of a sociotechnical system requires a framework for collecting and organizing data. One must know not only what information to gather, but how to put it into a form that is understandable. Based on extensive work carried out at the Tavistock Institute of Human Relations in London, England, an analytical model has been developed for this purpose (Foster, 1967). The model contains six steps which can be summarized as follows:

1. *Initial scanning:* The main characteristics of the work unit and its environment are identified in order to determine where the primary emphasis of the analysis needs to be placed.
2. *Identification of unit operations:* The principal stages in the production process are located and each identifiable segment is viewed as a unit operation with its own set of systemic properties.
3. *Identification of key process variances and their interrelationships:* The purpose of this step is to identify variances that arise from the production process or from the nature of the raw material. A variance refers to any deviation from some standard of

specification, and a variance is termed "key" if it significantly effects quantity or quality of production or operating or social costs.

4. *Analysis of the social system:* The main characteristics of the social system are examined and a table of variance control is constructed to determine the extent to which key variances are controlled by the social system. Ancillary activities, spatiotemporal relationships, job mobility, and payment systems are also examined, and work roles are tested against a list of basic psychological needs.
5. *Workers' perceptions of their roles:* An inquiry into how workers perceive their roles is made with special reference to the degree to which the basic psychological needs are met.
6. *Environmental analysis:* The environment of the work system is examined to ascertain how other units affect its functioning. More specifically, relations with the various support systems and the supply and user systems are studied. The main purpose is not to examine these units as sociotechnical systems, but to inspect the state of the relations between them and the experimental system. Finally, the work system is considered in the context of the larger organization with particular attention being paid to organization-wide developmental plans and general policies.

The process of applying the analytical model can be facilitated if some practical guidelines are followed. First, the experimental system should be examined in the state in which it is currently being operated. This provides a reality base for generating hypotheses for change and avoids the confusion of mixing up what is with what was or what could be. Second, data should be collected from a variety of sources. Workers, managers, and appropriate specialists offer different perceptual points which can be useful for determining system structure and functioning. Third, analysts often get into too much detail and as a consequence lose sight of the main problems of the system. A concerted effort should be made to list only key information and to avoid overelaboration of data. Paying particular attention to the initial scanning can help to eliminate unnecessary information. Fourth, results should be cross-checked for accuracy by allowing the data sources to view and discuss preliminary conclusions. In this respect, the analytical process should always be open to critical dialogue. Finally, hypotheses for redesign are often generated during the early stages of the analysis. These should be listed and carried forward into the next phase of the experiment.

Generating Hypotheses for Redesign

The primary outcome of the analytical phase of a sociotechnical experiment is to generate hypotheses for job and/or work-system redesign. Although an infinite variety of possible changes can be made, particular concern is for hypotheses aimed at joint optimization of the social and technological components and at a better matching between the work system and its environment.

Based on a number of job-related needs having to do with task variety, learning, autonomy, and social support and recognition, Emery (1963) has developed a preliminary set of work redesign principles. Since these principles need to be linked to the realities of an actual job if they are to have utility in a work context, they are best employed as general guidelines for generating specific hypotheses for redesign. Due to the importance of these principles, Emery is quoted in full.

At the level of the individual:

a. *Optimum variety of tasks within the job*. Too much variety can be inefficient for training and production as well as frustrating for the worker. However, too little can be conducive to boredom or fatigue. The optimum level would be that which allows the operator to take a rest from a high level of attention or effort or a demanding activity while working at another and, conversely, allow him to stretch himself and his capacities after a period of routine activity.

b. *A meaningful pattern of tasks that gives to each job a semblance of a single overall task*. The tasks should be such that although involving different levels of attention, degree of effort, or kinds of skill, they are interdependent: that is, carrying out one task makes it easier to get on with the next or gives a better end result to the overall task. Given such a pattern, the worker can help to find a method of working suitable to his requirements and can more easily relate his job to that of others.

c. *Optimum length of work cycle*. Too short a cycle means too much finishing and starting; too long a cycle makes it difficult to build up a rhythm of work.

d. *Some scope for setting standards of quantity and quality of production and a suitable feedback of knowledge of results*. Minimum standards generally have to be set by management to determine whether a worker is sufficiently trained, skilled or careful to hold the job. Workers are more likely to accept responsibility for higher standards if they have some freedom in setting them and are more likely to learn from the job if there is feedback. They can neither effectively set standards nor learn if there is not a quick enough feedback of knowledge of results.

e. *The inclusion in the job of some of the auxiliary and preparatory tasks*. The worker cannot and will not accept responsibility for matters outside his control. Insofar as the preceding criteria are met then the inclusion of such "boundary tasks" will extend the scope of the worker's responsibility and make for involvement in the job.

f. *The tasks included in the job should include some degree of care, skill, knowledge or effort that is worthy of respect in the community.*
g. *The job should make some perceivable contribution to the utility of the product for the consumer.*

At group level:

h. *Providing for "interlocking" tasks, job rotation or physical proximity where there is a necessary interdependence of jobs (for technical or psychological reasons).* At a minimum this helps to sustain communication and to create mutual understanding between workers whose tasks are interdependent and thus lessens friction, recriminations and "scape-goating." At best, this procedure will help to create work groups that enforce standards of cooperation and mutual help.
i. *Providing for interlocking tasks, job rotation or physical proximity where the individual jobs entail a relatively high degree of stress.* Stress can arise from apparently simple things such as physical activity, concentration, noise or isolation if these persist for long periods. Left to their own devices, people will become habituated, but the effects of the stress will tend to be reflected in more mistakes, accidents and the like. Communication with others in a similar plight tends to lessen the strain.
j. *Providing for interlocking tasks, job rotation or physical proximity where the individual jobs do not make an obvious perceivable contribution to the utility of the end product.*
k. *Where a number of jobs are linked together by interlocking tasks or job rotation they should as a group:*
   i. have some semblance of an overall task which makes a contribution to the utility of the product;
   ii. have some scope for setting standards and receiving knowledge of results:
   iii. have some control over the "boundary tasks."

Over extended social and temporal units:

1. *Providing for channels of communication so that the minimum requirements of the workers can be fed into the design of new jobs at an early stage.*
2. *Providing for channels of promotion to foreman rank which are sanctioned by the workers.*

Upon examining Emery's guides for work redesign, one can readily see that a particular redesign may lead beyond the individual job to group structures and other organizational units, including the structure of management. In examining the systemic properties of any work unit, the interdependent nature of task performance becomes a crucial issue. Changes in one part of the task network often amplify beyond the immediate work system to other parts of the organization. Although it may seem like a rudimentary issue, it is imperative that the primary and secondary effects—often unintended—of a redesign be examined in the larger context in which the experimental unit is embedded.

Creating redesign proposals involves a process of collecting and analyzing information and combining the data into new configurations to produce possible solutions. In short, one creates the new from the existing. Since such syntheses evolve from some knowledge of a problem, it is imperative that the analysis phase of the experiment provide a clear statement of the problems of the system. Once the designer is aware of the goals—social and economic—of the system, the difficulties to be overcome, the resources available, and the constraints which determine an acceptable solution, he or she can begin to generate solutions. Although we have listed guidelines for sociotechnical redesign, the actual generation of specific hypotheses is essentially a creative act. And as such, it is utimately constrained and consummated by the minds of individuals.

Testing and Evaluating Hypotheses for Redesign

Based on the hypotheses for redesign, one now proceeds to test and evaluate under experimental conditions. This phase of a sociotechnical experiment involves the following steps:

1. Reducing the list of possible hypotheses to a manageable set for experimental testing
2. Devising an action program for the testing period
3. Implementing the action program
4. Evaluating results

Since a variety of alternative hypotheses is often available for consideration, it is necessary to *reduce this list to a more manageable set for experimental testing*. Deciding which alternatives are most promising involves a judgment as to their viability against appropriate criteria. Although the actual criteria will vary from one context to another, they must eventually relate to the experimental system's objectives. In other words, they must pertain to production goals—quantity, quality, and operating costs—and to social objectives—meeting psychological needs and reducing stress, accidents, absenteeism, and turnover. Since some proposals deal exclusively with one type of objective while others involve a mix of both, it is imperative that the hypotheses be clearly identified as to their intended effects. And once these effects are identified, an assessment of the likelihood for achieving them must be made.

Given that one can ascertain the probability that a proposal can adequately produce results—at least at the subjective level—one must then deal with the issue of cost. That is to say, can the redesign achieve a desired outcome without undue social and economic costs? In determining cost effectiveness, it is necessary to project both short- and long-term consequences. For the social and economic effects of any proposal may take some time before they are actually manifested. One of the primary difficulties in determining costs is that many proposals produce outcomes which are multivalued. In short, individuals often place different values on the same result. For example, one hypothesis may reduce work hazards at the expense of productivity. Depending upon one's position—social responsibility versus economic rationality—the proposal will be evaluated differently. Although there is no optimal way to resolve a multivalued choice, the extent to which redesign proposals attempt to jointly optimize the social and technological systems can help to make this problem more manageable.

Once a decision is made concerning the hypotheses to be tested, a plan for implementing them under experimental conditions must be devised. An *action program* serves as a framework for introducing the changes in a systematic manner so that their impact can be properly evaluated. Since the structure of this plan is crucial for the implementation and evaluation stages of the experiment, it should include specific references to the following:

a. A detailed listing of the proposed changes
b. A timetable for the introduction of each change
c. A prospectus of the conditions for experimental protection
d. An inventory of extra services, tools, and materials needed for experimentation
e. A timetable for evaluative activities—instrumentation, analysis, and feedback
f. A clear listing of special rules or procedures that apply to the experimental period—job postings, vacations, sick leave, etc.
g. A determination of training needs and a program for meeting them
h. An account of supervisory functions and responsibilities during the experimental period

Inasmuch as these details go beyond the boundaries of the experimental system, the sanctioning body should take a decisive role in the creation of an action program. For without their active engagement and support, the conditions necessary for carrying out an experiment would be difficult if not impossible to attain.

Implementing a work-system change requires large amounts of patience and energy. Members of the experimental unit not only must learn the new, but must also learn to forget parts of the "old." Although this process varies according to the characteristics of the target system and of the change proposal, there are certain issues that are relevant to most sociotechnical change efforts. First, timing the change is critical for discontinuing the existing and starting the new. One should attempt to take advantage of naturally occurring disjunctions—the start of accounting periods, the end of vacations, the first of the month, the end of a preventive maintenance shutdown, and the change to a new product. These break points can serve as perceptual boundaries for workers, thereby enabling them to experience the existing cycle of events as coming to a natural completion before embarking on a new endeavor. This provides a sense of finishing an existing task in addition to furnishing a perceptual starting point for change. Second, focusing resources necessary for getting a change underway is important during the early stages of implementation. Providing the required tools, equipment, services, and other resources is essential for eliciting the synergy that is needed to get the ball rolling. There is nothing more damaging to a sociotechnical redesign than showing up for the kickoff and not having referees, a ball, and goalposts. Third, in carrying out social and technological change, one must be constantly aware that social change often lags technological change. The technological system can be implemented in one step, while the social system requires a series of intermediary stages as it develops towards maturity. A primary reason for this longer developmental process is that members of the social system must perceptually orient themselves to new conditions. Developing new role relationships and learning how to perform different tasks in novel ways often require a good deal of time. During the developmental stage, one can expect productivity to fall below the normal operating level and then oscillate up and down as workers modify their behaviors while experimenting with novel conditions. Although there is no adequate way of judging how long this process will take, the sociotechnical experimenter should be forewarned of this occurrence and consider it a normal part of the experimental process. Finally, changing an existing work unit frequently entails a shift in work-role definitions and relationships. Since existing roles are invested with a certain degree of status and emotionality, one can expect that anything which disrupts this pattern will be resisted. Instead of interpreting this phenomenon as resistance to change, one can view it as an active seeking toward positive affirmation. And, to the degree that a redesign negates existing status structures, one can anticipate that workers will actively seek to preserve those roles which provide them with positive affect. In lieu of worrying about resistance to change, one can see to it that new designs provide workers with increased opportunities for status and positive affirmation, thereby utilizing their positive motives on behalf of change.

All sociotechnical redesigns are hypotheses that must be *tested for their effectiveness* in the particular organizational setting in which they are employed. In this respect, each sociotechnical experiment is a case study which must be examined in its own milieu. In evaluating a redesign, it is necessary to deal with the following issues:

a. *The criterion problem*—on what grounds is one to measure the effectiveness of possible designs?
b. *The instrumentation problem*—can one validly and reliably measure the criteria?
c. *The research design problem*—is the research designed in such a way that one can draw valid inferences about the effects of the experimental treatment?

The criterion problem is concerned with standards for the evaluation of a redesign. Inasmuch as the evaluation of any redesign implies some form of hypothesis—either implicitly or explicitly formulated—which states that a specific design is expected to result in a particular outcome, we must relate the outcome to system goals. In short, criteria of effectiveness must reflect what we are trying to achieve—goal accomplishment. In deriving criteria, we are often confronted with the problem of choosing which aspects of the work system should serve as appropriate measures of performance. Since many objectives are multidimensional, one must frequently include several criteria that relate to the same goal. In addition, many economic objectives can be directly measured, but social goals frequently involve indirect measures. Therefore, in choosing measures of effectiveness, one should attempt to determine those variables—either direct or indirect—that bear on the system's objectives and include all or a representative sample of them as criteria.

The instrumentation problem involves the ability to measure the criteria of effectiveness. Once we have chosen appropriate criteria, we must then face the question: Can they be validly and reliably measured? In regard to instruments for data collection, validity refers to the ability of an instrument to measure what it purports to measure, while reliability refers to the consistency of the measures over time or across tests. The use of valid and reliable data collection techniques is critical in evaluating redesign experiments. If an instrument is not measuring what it claims to measure and/or if it is unstable in obtaining data, one has no assurance of what a redesign actually accomplished. In relation to many of the economic criteria, data collection appears fairly rudimentary. One collects pertinent information from existing organizational records with the assumption that it is valid and reliable. The collection of social data is often a more difficult issue than obtaining economic information. Standard techniques for measuring psychological need achievement, stress, alienation, involvement, satisfaction, learning, human development, and the like are often difficult to apply and interpret or unavailable. If one wishes to devise his or her own instruments, he or she is again faced with issues of validity and reliability. Although this is not meant to discourage a sociotechnical experimenter, it is imperative that one understand the importance of measuring criteria of effectiveness validly and reliably. And within these constraints, there is much room for innovation and ingenuity.

The research design problem is concerned with the question: Did the experimental redesign make a difference in this particular situation? The primary purpose of the testing and evaluation phase of an experiment is to examine the effects of possible redesigns. In doing this, each redesign is tested as an hypothesis for system change. In ascertaining the viability of any hypothesis, we are in fact rejecting alternative hypotheses. That is to say, in order to claim that a redesign had an effect on our experimental unit, one must show that extraneous variables—in the form of rival hypotheses—did not produce the results. These variables can best be controlled in a true experimental design in which the researcher has full control over the scheduling of the experimental treatment. Since few, if any, sociotechnical experiments fall into this category, one must then turn to a weaker yet valuable alternative—quasi-experimental designs (Campbell & Stanley, 1963). Quasi-experimental designs are appropriate when the experimenter has some control over the scheduling of data collection. Although they are not as strong as true experiments, quasi-experiments do allow one to control for some of the external variables. In this respect, they are useful where better designs are not feasible.

The importance of utilizing an appropriate research design in sociotechnical experimentation cannot be overstated. Given the amount of resources that are required for a typical experiment, it is important to design the research so that results can be adequately evaluated. Experimental design issues should be among the first problems resolved before any experimentation gets under way. This should allow full advantage to be taken over the control of scheduling of data collection, and it should alert the experimenter to possible research problems. In lieu of writing a text on sociotechnical experimental design, anyone contemplating a work-system experiment is strongly urged to thoroughly read Campbell and Stanley (1963) before attempting such an endeavor.

Transferring to a Normal Operating System

The transition from experimental testing to normal operating conditions involves a limited amount of experimental protection. Since the length of this handing-over process depends on the magnitude and type of change—larger deviations from the traditional taking longer—it is necessary to take these variables into account when determining the schedule for the reduction of protection. Inasmuch as the actual transition to a normal operating system takes place only after the unit has been operating successfully, it is imperative that evaluative activities continue for some time after the experimental stage. This not only provides information about the ability of the unit to remain in its new state, but also furnishes a data base for subsequent change. In carrying out continued evaluation, it is extremely useful to attempt to make this an ongoing feedback process. In effect, the evaluative criteria are used as indicators of sociotechnical functioning. To the extent that these indicators can be periodically employed and the information utilized by members of the work system, the experimental unit is provided with an ongoing critique of its operations and a built-in indicator of change.

Disseminating the Results

Once the experimental system is operating under normal working conditions, the process of disseminating results to other organizational units begins. While experience in diffusion of sociotechnical results is limited, two issues appear relevant to the diffusion process. The first is concerned with the *content* of dissemination. Deciding what to communicate to other organizational systems depends on the context of the experiment. If the conditions of the experimental site are similar to other work systems, the actual redesign or parts of it can be tried out in other units. Although these units will require some time for experimenting with the redesign, the analysis and hypothesis generation stages can be deleted at a great savings of time and energy. If the experimental site is different from other organizational systems, the process of carrying out a sociotechnical experiment can be communicated to interested work systems. Descriptions of the various stages of the experiment, advice on applying the strategy, redesign principles, and experimental designs are a few of the process dimensions that are useful for dissemination.

The second issue involves the *process* of dissemination. Since the diffusion process is essentially a communicative act, it is important to plan for how the results are to be communicated. Depending upon the particular audience—members of other work systems, managers, staff personnel, and members external to the organization—the variety of approaches for communicating results is almost unlimited. While the written word is certainly quite appealing, face-to-face strategies offer the advantage of two-way communication. Small discussion groups, informal site visits, and the like provide opportunities for interested individuals to clarify questions and to gain a better feel for the sociotechnical approach. In those cases where face-to-face communication is not feasible, videotape presentations by members of the experimental system can be extremely effective. In sum, the content and process of disseminating sociotechnical results are important factors in moving the approach to a larger organizational context; as such, they deserve considerable attention and planning.

CONCLUSIONS

Change, both within the organization and within the environment with which the organization interacts, is one of the most important dimensions of today's organizational life. Organizational members sometimes experience change as stimulating and productive but just as often as a source of confusion, tension, and frustration. One of the major determinates of how change is experienced is how it is introduced or managed by the organization. The intervention strategy reported in this paper is an attempt to manage social and technological change in a way that is both rewarding for the individual involved and productive for the organization. To provide the necessary protection and organizational sanction is to provide individuals with the opportunity to actively engage in exploring and experimenting with new ways of work. While the process is far from complete, this strategy is one attempt to help organizational members create and manage change.

REFERENCES

ARGYRIS, C. *The applicability of organizational sociology*. Cambridge University Press, 1972.

CAMPBELL, D. T. & STANLEY, J. C. *Experimental and quasi-experimental designs for research*. Chicago: Rand McNally, 1963.

CLARK, J. V. & KRONE, C. G. Towards an overall view of organizational development in the early seventies. In J. M. Thomas and W. G. Bennis (Eds.) *The management of change and conflict*. Penguin Books, 1972.

DAVIS, L. E. & CANTER, R. R. Job design research. *Journal of Industrial Engineering*, 1956, *VII*, 275.

DAVIS, L. E. & TRIST, E. L. Improving the quality of work life: Experience of the socio-technical approach. *Work in America*. U.S. Department of Health, Education, and Welfare, 1972.

EMERY, F. E. Some hypotheses about the way in which tasks may be more effectively put together to make jobs. London: Tavistock Institute of Human Relations, Doc. No. T.176, 1963.

FOSTER, M. Developing an analytical model for socio-technical analysis. London: Tavistock Institute of Human Relations, Doc. No. HRC7, 1967.

RICE, A. K. *Productivity and social organization: The Ahmedabad experiment*. London: Tavistock Publications, 1958.

THORSUND, E. Industrial democracy. London: Tavistock Institute of Human Relations, Doc. No. T.886, 1966.

TRIST, E. L. & BAMFORTH, K. N. Some social and psychological consequences of the longwall method of coal-getting. *Human Relations*, 1951, *4*, 2-38.

TRIST, E. L., HIGGIN, G. W., MURRAY, H. & POLLACK, A. B. *Organizational choice*. London: Tavistock Publications, 1963.

How to Counter Alienation in the Plant

Richard E. Walton*

Managers don't need anyone to tell them that employee alienation exists. Terms such as "blue-collar blues" and "salaried drop-outs" are all too familiar. But are they willing to undertake the major innovations necessary for redesigning work organizations to deal effectively with the root causes of alienation? My purpose in this article is to urge them to do so, for two reasons:

1. The current alienation is not merely a phase that will pass in due time.

2. The innovations needed to correct the problem can simultaneously enhance the quality of work life (thereby lessening alienation) and improve productivity.

In the first part of the article, I shall risk covering terrain already familiar to some readers in order to establish that alienation is a basic, longterm, and mounting problem. Then I shall present some examples of the comprehensive redesign that I believe is required.

I also hope to provide today's managers with a glimpse at what may be the industrial work environment of the future, as illustrated by a pet-food plant which opened in January 1971.

Author's note: An earlier version of this article was prepared for the Work in America Project, sponsored by the Secretary of the Department of Health, Education and Welfare, as a basis for assessing the nature of problems and potential crises associated with work in the United States.

*Edsel Bryant Ford professor of business administration and director of the Division of Research, Harvard Business School. From *Harvard Business Review*, November-December, 1972, pp. 70-81.

In this facility, management set out to incorporate features that would provide a high quality work life, enlist unusual human involvement, and result in high productivity. The positive results of the experiment to date are impressive, and the difficulties encountered in implementing it are instructive. Moreover, similar possibilities for *comprehensive* innovations exist in a wide variety of settings and industries.

The word "comprehensive" is important because my argument is that each technique in the standard fare of personnel and organization development programs (e.g., job enrichment, management by objectives, sensitivity training, confrontation and team-building sessions, participative decision making) has grasped only a limited truth and has fallen far short of producing meaningful change. In short, more radical, comprehensive, and systemic redesign of organizations is necessary.

ANATOMY OF ALIENATION

There are two parts to the problem of employee alienation: (1) the productivity output of work systems, and (2) the social costs associated with employee inputs. Regarding the first, U.S. productivity is not adequate to the challenges posed by international competition and inflation; it cannot sustain impressive economic growth. (I do not refer here to economic growth as something to be valued merely for its own sake—it is politically a precondition for the income redistribution that will make equality of opportunity possible in the United States.) Regarding the second, the social and psychological costs of work systems are excessive, as evidenced by their effects on the mental and physical health of employees and on the social health of families and communities.

Employee alienation *affects* productivity and *reflects* social costs incurred in the workplace. Increasingly, blue- and white-collar employees and, to some extent, middle managers tend to dislike their jobs and resent their bosses. Workers tend to rebel against their union leaders. They are becoming less concerned about the quality of the product of their labor and more angered about the quality of the context in which they labor.

In some cases, alienation is expressed by passive withdrawal—tardiness, absenteeism and turnover, and inattention on the job. In other cases, it is expressed by active attacks—pilferage, sabotage, deliberate waste, assaults, bomb threats, and other disruptions of work routines. Demonstrations have taken place and underground newspapers have appeared in large organizations in recent years to protest company policies. Even more recently, employees have cooperated with newsmen, Congressional committees, regulatory agencies, and protest groups in exposing objectionable practices.

These trends all have been mentioned in the media, but one expression of alienation has been underreported: pilferage and violence against property and persons. Such acts are less likely to be revealed to the police and the media when they occur in a private company than when they occur in a high school, a ghetto business district, or a suburban town. Moreover, dramatic increases in these forms of violence are taking place at the plant level. This trend is not reported in local newspapers and there is little or no appreciation of it at corporate headquarters. Local management keeps quiet because violence is felt to reflect unfavorably both on its effectiveness and on its plant as a place to work.

Roots of Conflict

The acts of sabotage and other forms of protest are overt manifestations of a conflict between changing employee attitudes and organizational inertia. Increasingly, what employees expect from their jobs is different from what organizations are prepared to offer them. These evolving expectations of workers conflict with the demands, conditions, and rewards of employing organizations in at least six important ways:

1. Employees want challenge and personal growth, but work tends to be simplified and specialties tend to be used repeatedly in work assignments. This pattern exploits the narrow skills of a worker, while limiting his or her opportunities to broaden or develop.

2. Employees want to be included in patterns of mutual influence; they want egalitarian treatment. But organizations are characterized by tall hierarchies, status differentials, and chains of command.

3. Employee commitment to an organization is increasingly influenced by the intrinsic interest of the work itself, the human dignity afforded by management, and the social responsibility reflected in the organization's products. Yet organization practices still emphasize material rewards and employment security and neglect other employee concerns.

4. What employees want from careers, they are apt to want *right now*. But when organizations design job hierarchies and career paths, they continue to assume that today's workers are as willing to postpone gratifications as were yesterday's workers.

5. Employees want more attention to the emotional aspects of organization life, such as individual self-esteem, openness between people, and expressions of warmth. Yet organizations emphasize rationality and seldom legitimize the emotional part of the organizational experience.

6. Employees are becoming less driven by competitive urges, less likely to identify competition as the "American way." Nevertheless, managers continue to plan career patterns, organize work, and design reward systems as if employees valued competition as highly as they used to.

Pervasive Social Forces. The foregoing needs and desires that employees bring to their work are but a local reflection of more basic, and not readily reversible, trends in U.S. society. These trends are fueled by family and social experience as well as by social institutions, especially schools. Among the most significant are:

The rising level of education—Employees bring to the workplace more abilities and, correspondingly, higher expectations than in the past.

The rising level of wealth and security—Vast segments of today's society never have wanted for the tangible essentials of life; thus they are decreasingly motivated by pay and security, which are taken for granted.

The decreased emphasis given by churches, schools, and families to obedience to authority—These socialization agencies have promoted individual initiative, self-responsibility and -control, the relativity of values, and other social patterns that make subordinacy in traditional organizations an increasingly bitter pill to swallow for each successive wave of entrants to the U.S. work force.

The decline in achievement motivation—For example, whereas the books my parents read in primary school taught them the virtues of hard work and competition, my children's books emphasize self-expression and actualizing one's potential. The workplace has not yet fully recognized this change in employee values.

The shifting emphasis from individualism to social commitment—This shift is driven in part by a need for the direct gratifications of human connectedness (for example, as provided by commune living experiments). It also results from a growing appreciation of our interdependence, and it renders obsolete many traditional workplace concepts regarding the division of labor and work incentives.

Exhibit 1 shows how these basic societal forces underlie, and contribute to, the problem of alienation and also sums up the discussion thus far. Actually, I believe that protests in the workplace will mount even more rapidly than is indicated by the contributing trends postulated here. The latent dissatisfaction of workers will be activated as (a) the issues receive public attention and (b) some examples of attempted solutions serve to raise expectations (just as the blacks' expressions of dissatisfaction with

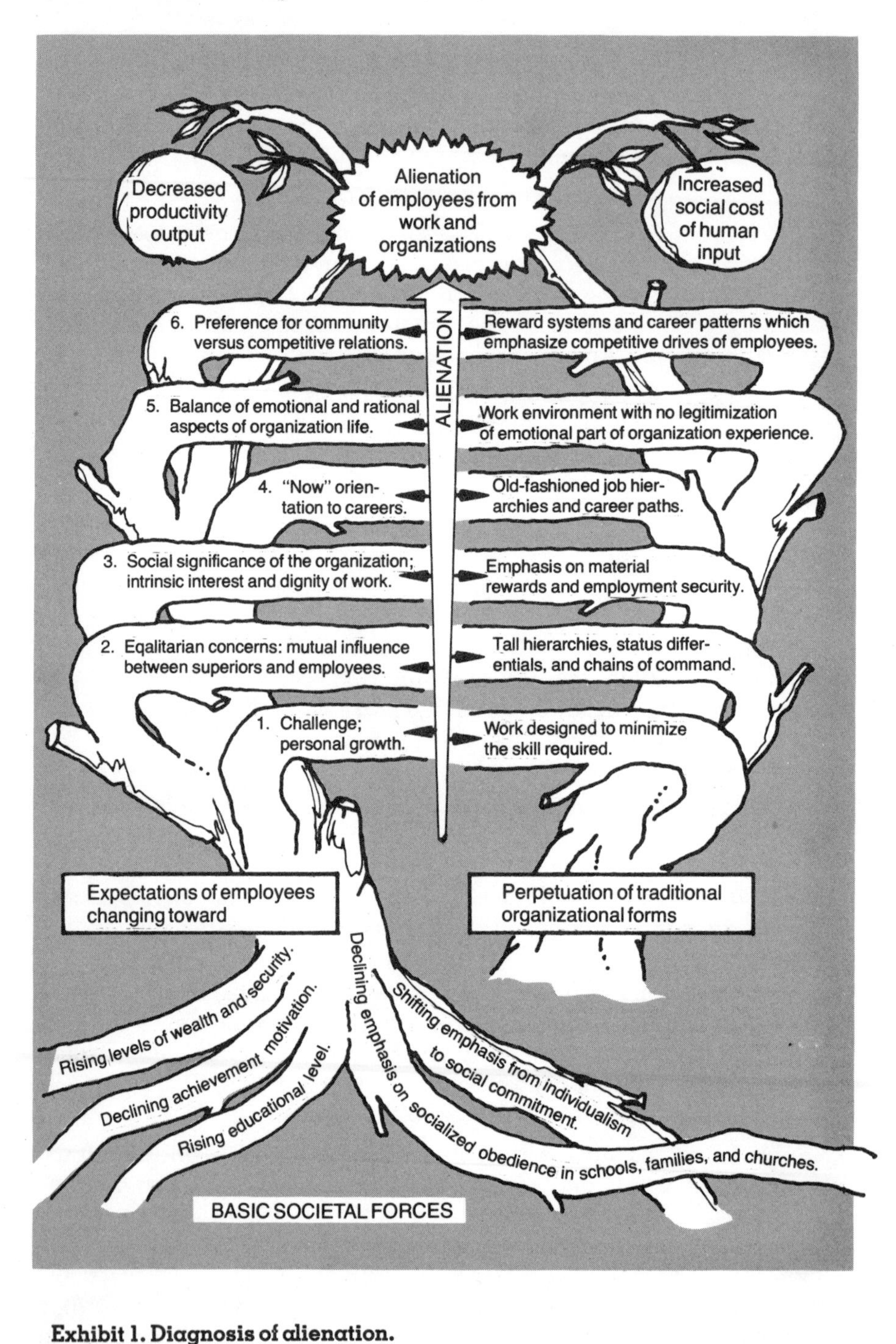

Exhibit 1. Diagnosis of alienation.

social and economic inequities were triggered in the 1950's, and women's discontent expanded late in the 1960's).

Revitalization and Reform

It seems clear that employee expectations are not likely to revert to those of an earlier day. As *Exhibit 1* shows, the conflicts between these expectations and traditional organizations result in alienation. This alienation, in turn, exacts a deplorable psychological and social cost as well as causing worker behavior that depresses productivity and constrains growth. In short, we need major innovative efforts to redesign work organizations, efforts that take employee expectations into account.

Over the past two decades we have witnessed a parade of organization development, personnel, and labor relations programs that promised to revitalize organizations:

Job enrichment would provide more varied and challenging content in the work.

Participative decision making would enable the information, judgments, and concerns of subordinates to influence the decisions that affect them.

Management by objectives would enable subordinates to understand and shape the objectives toward which they strive and against which they are evaluated.

Sensitivity training or encounter groups would enable people to relate to each other as human beings with feelings and psychological needs.

Productivity bargaining would revise work rules and increase management's flexibility with a quid pro quo whereby the union ensures that workers share in the fruits of the resulting productivity increases.

Each of the preceding programs *by itself* is an inadequate reform of the workplace and has typically failed in its more limited objectives. While application is often based in a correct diagnosis, each approach is only a partial remedy; therefore, the organizational system soon returns to an earlier equilibrium.

The lesson we must learn in the area of work reform is similar to one we have learned in another area of national concern. It is now recognized that a health program, a welfare program, a housing program, or an employment program alone is unable to make a lasting impact on the urban-poor syndrome. Poor health, unemployment, and other interdependent aspects of poverty must be attacked in a coordinated or systemic way.

So it is with meaningful reform of the workplace: we must think "systemically" when approaching the problem. We must coordinate the redesign of the ways tasks are packaged into jobs, the way workers are required to relate to each other, the way performance is measured and rewards are made available, the way positions of authority and status symbols are structured, and the way career paths are conceived. Moreover, because these types of changes in work organizations imply new employee skills and different organizational cultures, transitional programs must be established.

A PROTOTYPE OF CHANGE

A number of major organization design efforts meet the requirements of being systemic and comprehensive. One experience in which I have been deeply involved is particularly instructive. As a recent and radical effort, it generally encompasses and goes beyond what has been done elsewhere.

During 1968, a large pet-food manufacturer was planning an additional plant at a new location. The existing manufacturing facility was then experiencing many of the symptoms of alienation that I have already outlined. There were frequent instances of employee indifference and inattention that, because of the continuous-process technology, led to plant shutdowns, product waste, and costly recycling. Employees effectively worked only a modest number of hours per day, and they resisted changes toward fuller utilization of manpower. A series of acts of sabotage and violence occurred.

Because of these pressures and the fact that it was not difficult to link substantial manufacturing costs to worker alienation, management was receptive to basic innovations in the new plant. It decided to design the plant to both accommodate changes in the expectations of employees and utilize knowledge developed by the behavioral sciences.

Key Design Features

The early development of the plant took more than two years. This involved planning, education, skill training, and building the nucleus of the new organization into a team.

During this early period, four newly selected managers and their superior met with behavioral science experts and visited other industrial plants that were experimenting with innovative organizational methods. Thus they were stimulated to think about departures from traditional work organizations and given reassurance that other organizational modes were not only possible but also more viable in the current social context. While the consultations and plant visits provided some raw material for designing the new organization, the theretofore latent knowledge of the five managers played the largest role. Their insights into the aspirations of people and basically optimistic assumptions about the capacities of human beings were particularly instrumental in the design of the innovative plant. In the remainder of this section, I shall present the nine key features of this design.

1. *Autonomous Work Groups*. Self-managed work teams are given collective responsibility for large segments of the production process. The total work force of approximately 70 employees is organized into six teams. A processing team and a packaging team operate during each shift. The processing team's jurisdiction includes unloading, storage of materials, drawing ingredients from storage, mixing, and then performing the series of steps that transform ingredients into a pet-food product. The packaging team's responsibilities include the finishing stages of product manufacturing—packaging operations, warehousing, and shipping.

A team is comprised of from 7 to 14 members (called "operators") and a team leader. Its size is large enough to include a natural set of highly interdependent tasks, yet small enough to allow effective face-to-face meetings for decision making and coordination. Assignments of individuals to sets of tasks are subject to team consensus. Although at any given time one operator has primary responsibility for a set of tasks within the team's jurisdiction, some tasks can be shared by several operators. Moreover, tasks can be redefined by the team in light of individual capabilities and interests. In contrast, individuals in the old plant were permanently assigned to specific jobs.

Other matters that fall within the scope of team deliberation, recommendation, or decision making include:

- Coping with manufacturing problems that occur within or between the teams' areas of responsibilities.
- Temporarily redistributing tasks to cover for absent employees.
- Selecting team operators to serve on plant-wide committees or task forces.
- Screening and selecting employees to replace departing operators.
- Counseling those who do not meet team standards (e.g., regarding absences or giving assistance to others).

2. *Integrated Support Functions*. Staff units and job specialties are avoided. Activities typically performed by maintenance, quality control, custodial, industrial engineering, and personnel units are built into

an operating team's responsibilities. For example, each team member maintains the equipment he operates (except for complicated electrical maintenance) and housekeeps the area in which he works. Each team has responsibility for performing quality tests and ensuring quality standards. In addition, team members perform what is normally a personnel function when they screen job applicants.

3. *Challenging Job Assignments*. While the designers understood that job assignments would undergo redefinition in light of experience and the varying interests and abilities on the work teams, the initial job assignments established an important design principle. Every set of tasks is designed to include functions requiring higher-order human abilities and responsibilities, such as planning, diagnosing mechanical or process problems, and liaison work.

The integrated support functions just discussed provide one important source of tasks to enrich jobs. In addition, the basic technology employed in the plant is designed to eliminate dull or routine jobs as much as possible. But some nonchallenging, yet basic, tasks still have to be compensated for. The forklift truck operation, for example, is not technically challenging. Therefore, the team member responsible for it is assigned other, more mentally demanding tasks (e.g., planning warehouse space utilization and shipping activities).

Housekeeping duties are also included in every assignment, despite the fact that they contribute nothing to enriching the work, in order to avoid having members of the plant community who do nothing but menial cleaning.

4. *Job Mobility and Rewards for Learning*. Because all sets of tasks (jobs) are designed to be equally challenging (although each set comprises unique skill demands), it is possible to have a single job classification for all operators. Pay increases are geared to an employee mastering an increasing proportion of jobs first in the team and then in the total plant. In effect, team members are payed for learning more and more aspects of the total manufacturing system. Because there are no limits on the number of operators that can qualify for higher pay brackets, employees are also encouraged to teach each other. The old plant, in contrast, featured large numbers of differentiated jobs and numerous job classifications, with pay increases based on progress up the job hierarchy.

5. *Facilitative Leadership*. Team leaders are chosen from foreman-level talent and are largely responsible for team development and group decision making. This contrasts with the old plant's use of supervisors to plan, direct, and control the work of subordinates. Man-

agement feels that in time the teams will be self-directed and so the formal team leader position might not be required.

6. *"Managerial" Decision Information for Operators*. The design of the new plant provides operators with economic information and managerial decision rules. Thus production decisions ordinarily made by supervisors can now be made at the operator level.

7. *Self-Government for the Plant Community*. The management group that developed the basic organization plan before the plant was manned refrained from specifying in advance any plant rules. Rather, it is committed to letting these rules evolve from collective experience.

8. *Congruent Physical and Social Context*. The differential status symbols that characterize traditional work organizations are minimized in the new plant. There is an open parking lot, a single entrance for both the office and plant, and a common decor throughout the reception area, offices, locker rooms, and cafeteria.

The architecture facilitates the congregating of team members during working hours. For example, rather than following the plan that made the air conditioned control room in the process tower so small that employees could not congregate there, management decided to enlarge it so that process team operators could use it when not on duty elsewhere. The assumption here is that rooms which encourage ad hoc gatherings provide opportunities not only for enjoyable human exchanges but also for work coordination and learning about others' jobs.

9. *Learning and Evolution*. The most basic feature of the new plant system is management's commitment to continually assess both the plant's productivity and its relevance to employee concerns in light of experience.

I believe pressures will mount in this system with two apparently opposite implications for automation:

- On the one hand, people will consider ways of automating the highly repetitive tasks. (There are still back-breaking routine tasks in this plant; for example, as 50-pound bags pile up at the end of the production line, someone must grab them and throw them on a pallet.)
- On the other hand, some processes may be slightly de-automated. The original design featured fully automated or "goof-proof" systems to monitor and adjust several segments of the manufacturing process; yet some employees have become confident that they can improve on the systems if they are allowed to intervene with their own judgments. These employees suggest that organizations may benefit more from operators who are alert and who care than from goof-proof systems.

Implementation Difficulties

Since the plant start-up in January 1971, a number of difficulties have created at least temporary, and in some cases enduring, gaps between ideal expectations and reality.

The matter of compensation, for example, has been an important source of tension within this work community. There are four basic pay rates: starting rate, single job rate (for mastering the first job assignment), team rate (for mastering all jobs within the team's jurisdiction), and plant rate. In addition, an employee can qualify for a "specialty" add-on if he has particular strengths—e.g., in electrical maintenance.

Employees who comprised the initial work force were all hired at the same time, a circumstance that enabled them to directly compare their experiences. With one or two exceptions on each team, operators all received their single job rates at the same time, about six weeks after the plant started. Five months later, however, about one third of the members of each team had been awarded the team rate.

The evaluative implications of awarding different rates of pay have stirred strong emotions in people who work so closely with each other. The individual pay decisions had been largely those of the team leaders who, however, were also aware of operators' assessments of each other. In fact, pay rates and member contributions were discussed openly between team leaders and their operators as well as among operators themselves. Questions naturally arose:

- Were the judgments about job mastery appropriate?
- Did everyone have an equal opportunity to learn other jobs?
- Did team leaders depart from job mastery criteria and include additional considerations in their promotions to team rate?

Thus the basic concepts of pay progression are not easy to treat operationally. Moreover, two underlying orientations compete with each other and create ambivalences for team leaders and operators alike:

- A desire for more equality, which tends to enhance cohesiveness.
- A desire for more differential rewards for individual merit, which may be more equitable but can be divisive.

Similar team and operator problems have also occurred in other areas. Four of these are particularly instructive and are listed in the ruled insert at the end of this article.

Management, too, has been a source of difficulty. For example, acceptance and support from superiors and influential staff groups at corporate headquarters did not always come easily, thus creating anxiety and uncertainty within the new plant community.

Management resistance to innovative efforts of this type has a variety of explanations apart from natural and healthy skepticism. Some staff departments feel threatened by an experiment in which their functions no longer require separate units at the plant level. Other headquarters staff who are not basically threatened may nevertheless resist an innovation that deviates from otherwise uniform practices in quality control, accounting, engineering, or personnel. Moreoever, many managers resent radical change, presuming that it implies they have been doing their jobs poorly.

Evidence of Success

While the productivity and the human benefits of this innovative organization cannot be calculated precisely, there have nevertheless been some impressive results:

- Using standard principles, industrial engineers originally estimated that 110 employees should man the plant. Yet the team concept, coupled with the integration of support activities into team responsibilities, has resulted in a manpower level of slightly less than 70 people.
- After 18 months, the new plant's fixed overhead rate was 33% lower than in the old plant. Reductions in variable manufacturing costs (e.g., 92% fewer quality rejects and an absenteeism rate 9% below the industry norm) resulted in annual savings of $600,000. The safety record was one of the best in the company and the turnover was far below average. New equipment is responsible for some of these results, but I believe that more than one half of them derive from the innovative human organization.
- Operators, team leaders, and managers alike have become more involved in their work and also have derived high satisfaction from it. For example, when asked what work is like in the plant and how it differs from other places they have worked, employees typically replied: "I never get bored." "I can make my own decisions." "People will help you; even the operations manager will pitch in to help you clean up a mess—he doesn't act like he is better than you are." I was especially impressed with the diversity of employees who made such responses. Different operators emphasized different aspects of the work culture, indicating that the new system had unique meaning for each member. This fact confirms the importance of systemwide innovation. A program of job enrichment, for example, will meet the priority psychological needs of one worker, but not another. Other single efforts are similarly limited.

• Positive assessments of team members and team leaders in the new plant are typically reciprocal. Operators report favorably on the greater influence that they enjoy and the open relations which they experience between superiors and themselves; superiors report favorably on the capacities and sense of responsibility that operators have developed.

• While the plant is not without the occasional rumor that reflects some distrust and cynicism, such symptomatic occurrences are both shorter-lived and less frequent than are those that characterize other work organizations with which I am familiar. Similarly, although the plant work force is not without evidence of individual prejudice toward racial groups and women, I believe that the manifestations of these social ills can be handled more effectively in the innovative environment.

• Team leaders and other plant managers have been unusually active in civic affairs (more active than employees of other plants in the same community). This fact lends support to the theory that participatory democracy introduced in the plant will spread to other institutional settings. Some social scientists, notably Carole Pateman, argue that this will indeed be the case.[1]

• The apparent effectiveness of the new plant organization has caught the attention of top management and encouraged it to create a new corporate-level unit to transfer the organizational and managerial innovations to other work environments. The line manager responsible for manufacturing, who initiated the design of the innovative system, was chosen to head this corporate diffusion effort. He can now report significant successes in the organizational experiments under way in several units of the old pet-food plant.

What It Cost

I have already suggested what the pet-food manufacturer expected to gain from the new plant system: a more reliable, more flexible, and lower-cost manufacturing plant; a healthier work climate; and learning that could be transferred to other corporate units.

What did it invest? To my knowledge, no one has calculated the extra costs incurred prior to and during start-up that were specifically related to the innovative character of the organization. (This is probably because such costs were relatively minor compared with the amounts involved in other decisions made during the same time period.) How-

---

[1]*Participation and Democratic Theory* (Cambridge, England, Cambridge University Press, 1970).

ever, some areas of extra cost can be cited:

Four managers and six team leaders were brought on board several months earlier than they otherwise would have been. The cost of outside plant visits, training, and consulting was directly related to the innovative effort. And a few plant layout and equipment design changes, which slightly increased the initial cost of the new plant, were justified primarily in terms of the organizational requirements.

During the start-up of the new plant, there was a greater than usual commitment to learning from doing. Operators were allowed to make more decisions on their own and to learn from their own experience, including mistakes. From my knowledge of the situation, I infer that there was a short-term—first quarter—sacrifice of volume, but that it was recouped during the third quarter when the more indelible experiences began to pay off. In fact, I would be surprised if the pay-back period for the company's entire extra investment was greater than the first year of operation.

Why It Works

Listed in the ruled insert on page 207 are eight factors that influenced the success of the new pet-food plant. I want to stress, however, that these are merely facilitating factors and are *not* preconditions for success.

For example, while a new plant clearly facilitates the planning for comprehensive plantwide change (Factor 3), such change is also possible in ongoing plants. In the latter case, the change effort must focus on a limited part of the plant—say, one department or section at a time. Thus, in the ongoing facility, one must be satisfied with a longer time horizon for plantwide innovation.

Similarly, the presence of a labor union (Factor 6) does not preclude innovation, although it can complicate the process of introducing change. To avoid this, management can enter into a dialogue with the union about the changing expectations of workers, the need for change, and the nature and intent of the changes contemplated. Out of such dialogue can come an agreement between management and union representatives on principles for sharing the fruits of any productivity increases.

One factor I do regard as essential, however, is that the management group immediately involved must be committed to innovation and able to reach consensus about the guiding philosophy for the organization. A higher-level executive who has sufficient confidence in the innovative effort is another essential. He or she will act to protect the experiment from premature evaluations and from the inevitable, reactive pressures to bring it into line with existing corporate policies and practices.

Management and supervisors must work hard to make such a system succeed—harder, I believe, than in a more traditional system. In the case of the pet-food group, more work was required than in the traditional plant, but the human satisfactions were also much greater.

THE OTHER INNOVATORS

While the pet-food plant has a unique character and identity, it also has much in common with innovative plants of such U.S. corporations as Procter & Gamble and TRW Systems. Moreover, innovative efforts have been mounted by many foreign-based companies—e.g., Shell Refining Co., Ltd. (England), Northern Electric Co., Ltd. (Canada), Alcan Aluminum (smelting plants in Quebec Province, Canada), and Norsk-Hydro (a Norwegian manufacturer of fertilizers and chemicals). Related experiments have been made in the shipping industry in Scandinavia and the textile industry in Ahmedabad, India. Productivity increases or benefits for these organizations are reported in the range of 20% to 40% and higher, although I should caution that all evidence on this score involves judgment and interpretation.

All of these experiments have been influenced by the pioneering effort made in 1950 in the British coal mining industry by Eric Trist and his Tavistock Institute colleagues.[2]

Procter & Gamble has been a particularly noteworthy innovator. One of its newer plants includes many design features also employed in the pet-food plant. High emphasis has been placed on the development of "business teams" in which organization and employee identification coincides with a particular product family. Moreover, the designers were perhaps even more ambitious than their pet-food predecessors in eliminating first-line supervision. In terms of performance, results are reportedly extraordinary, although they have not been publicized. In addition, employees have been unusually active in working for social change in the outside community.[3]

Progressive Assembly Lines

Critics often argue that experiments like those I have discussed are not transferable to other work settings, especially ones that debase human

---

[2]See E. L. Trist, G. W. Higgin, H. Murray, and A. B. Pollock, *Organizational Choice* (London, Tavistock Publications, 1963).

[3]Personal correspondence with Charles Krone, Internal Consultant, Procter & Gamble.

dignity. The automobile assembly line is usually cited as a case in point.

I agree that different work technologies create different opportunities and different levels of constraint. I also agree that the automotive assembly plant represents a difficult challenge to those who wish to redesign work to decrease human and social costs and increase productivity. Yet serious experimental efforts to meet these challenges are now under way both in the United States and overseas.

To my knowledge, the most advanced projects are taking place in the Saab-Scandia automotive plants in Södertälje, Sweden. Consider, for example, these major design features of a truck assembly plant:

- Production workers have been included as members of development groups that discuss such matters as new tool and machine designs before they are approved for construction.
- Workers leave their stations on the assembly line for temporary assignments (e.g., to work with a team of production engineers "rebalancing" jobs on the line).
- Responsibility for in-process inspection has been shifted from a separate quality-inspection unit to individual production workers. The separate quality section instead devotes all its efforts to checking and testing completed trucks.
- Work tasks have been expanded to include maintenance care of equipment, which was previously the responsibility of special mechanics.
- Individuals have been encouraged to learn several jobs. In some cases, a worker has proved capable of assembling a complete engine.

Encouraged by the results of these limited innovations, the company is applying them in a new factory for the manufacture and assembly of car engines, which was opened in January 1972. In the new plant, seven assembly groups have replaced the continuous production line; assembly work within each group is not controlled mechanically; and eventually the degree of specialization, methods of instruction, and work supervision will vary widely among the assembly groups.

In effect, the seven groups fall along a spectrum of decreasing specialization. At one end is a group of workers with little or no experience in engine assembly; at the other end is a group of workers with extensive experience in total engine assembly. It is hoped that, ultimately, each group member will have the opportunity to assemble an entire engine.[4]

---

[4]For a more complete description of this plant, see Jan-Peter Norstedt, *Work Organization and Job Design at Saab-Scandia in Södertälje* (Stockholm, Technical Department, Swedish Employers' Confederation, December 1970).

In addition to the improvements that have made jobs more interesting and challenging for workers, management anticipates business gains that include: (a) a work system less sensitive to disruption than is the production line (a factor of considerable significance in the company's recent experience); and (b) the twofold ability to recruit workers and reduce absenteeism and turnover. (The company has encountered difficulty in recruiting labor and has experienced high turnover and absenteeism.)

Another Swedish company, Volvo, also has ambitious programs for new forms of work systems and organization. Especially interesting is a new type of car assembly plant being built at Kalmar. Here are its major features:

- Instead of the traditional assembly line, work teams of 15-25 men will be assigned responsibility for particular sections of a car (e.g., the electrical system, brakes and wheels, steering and controls).
- Within teams, members will decide how work should be divided and distributed.
- Car bodies will be carried on self-propelled carriages controlled by the teams.
- Buffer stocks between work regions will allow variations in the rate of work and "stock piling" for short pauses in the work flow.
- The unique design of the building will provide more outside windows, many small workshops to reinforce the team atmosphere, and individual team entrances, changing rooms, and relaxation areas.

The plant, scheduled to open in 1974, will cost 10% more than a comparable conventional car plant, or an estimated premium of $2 million. It will employ 600 people and have a capacity to produce 30,000 cars each year. Acknowledging the additional capital investment per employee, with its implication for fixed costs, Volvo nevertheless justified this experiment as "another stage in the company's general attempt to create greater satisfaction at work."[5]

Question of Values

The designers of the Procter and Gamble and pet-food plants were able to create organizational systems that both improved productivity and enhanced the quality of work life for employees. It is hard to say, however, whether the new Saab-Scandia and Volvo plants will result in comparable improvements in both areas. (As I mentioned earlier, the assembly line presents a particularly difficult challenge.)

[5]Press release from Volvo offices, Gothenburg, Sweden, June 29, 1972.

In any event, I am certain that managers who concern themselves with these two values will find points at which they must make trade-offs—i.e., that they can only enhance the quality of work life at the expense of productivity or vice versa. What concerns me is that it is easier to measure productivity than to measure the quality of work life, and that this fact will bias how trade-off situations are resolved.

Productivity may not be susceptible to a single definition or to precise measurement, but business managers do have ways of gauging changes in it over time and comparing it from one plant to the next. They certainly can tell whether their productivity is adequate for their competitive situation.

But we do not have equally effective means for assessing the quality of work life or measuring the associated psychological and social costs and gains for workers.[6] We need such measurements if this value is to take its appropriate place in work organizations.

CONCLUSION

The emerging obligation of employers in our society is a twin one: (1) to use effectively the capacities of a major natural resource—namely, the manpower they employ; and (2) to take steps to both minimize the social costs associated with utilizing that manpower and enhance the work environment for those they employ.

Fulfillment of this obligation requires major reform and innovation in work organizations. The initiative will eventually come from many quarters, but I urge professional managers and professional schools to take leadership roles. There are ample behavioral science findings and a number of specific experiences from which to learn and on which to build.

Furthermore, the nature of the problem and the accumulating knowledge about solutions indicate that organizational redesign should be systemic; it should embrace the division of labor, authority and status structures, control procedures, career paths, allocation of the economic fruits of work, and the nature of social contacts among workers. Obviously, the revisions in these many elements must be coordinated and must result in a new, internally consistent whole.

This call for widespread innovation does *not* mean general application of a particular work system, such as the one devised for the pet-food

---

[6]For the beginning of a remedy to this operational deficiency, see Louis E. Davis and Eric L. Trist, *Improving the Quality of Work Life: Experience of the Socio-Technical Approach* (Washington, D.C., Upjohn Institute, scheduled for publication at the time this article was written).

plant. There are important differences within work forces and between organizations. Regional variances, education, age, sex, ethnic background, attitudes developed from earlier work experiences, and the urban-rural nature of the population all will influence the salient expectations in the workplace. Moreover, there are inherent differences in the nature of primary task technologies, differences that create opportunities for and impose constraints on the way work can be redesigned.

Implementation problems in the pet-food plant

Here are four team and operator problems encountered in the design of the innovative plant:

1 The expectations of a small minority of employees did not coincide with the demands placed on them by the new plant community. These employees did not get involved in the spirit of the plant organization, participate in the spontaneous mutual-help patterns, feel comfortable in group meetings, or appear ready to accept broader responsibilities. For example, one employee refused to work in the government-regulated product-testing laboratory because of the high level of responsibility inherent in that assignment.

2 Some team leaders have had considerable difficulty *not* behaving like traditional authority figures. Similarly, some employees have tried to elicit and reinforce more traditional supervisory patterns. In brief, the actual expectations and preferences of employees in this plant fall on a spectrum running from practices idealized by the system planners to practices that are typical of traditional industrial plants. They do, however, cluster toward the idealized end of the spectrum.

3 The self-managing work teams were expected to evolve norms covering various aspects of work, including responsible patterns of behavior (such as mutual help and notification regarding absences). On a few occasions, however, there was excessive peer group pressure for an individual to conform to group norms.

Scapegoating by a powerful peer group is as devastating as scapegoating by a boss. The same is true of making arbitrary judgments. Groups, however, contain more potential for checks and balances, understanding and compassion, reason and justice. Hence it is important for team leaders to facilitate the development of these qualities in work groups.

4 Team members have been given assignments that were usually limited to supervisors, managers, or professionals: heading the plant safety committee, dealing with outside vendors, screening and selecting new employees, and traveling to learn how a production problem is handled in another plant or to trouble-shoot a shipping problem. These assignments have been heady experiences for the operators, but have also generated mixed feelings among others. For example, a vendor was at least initially disappointed to be dealing with a worker because he judged himself in part by his ability to get to higher organizational levels of the potential customer (since typically that is where decisions are made). In another case, a plant worker attended a corporationwide meeting of safety officials where all other representatives were from management. The presence and implied equal status of the articulate, knowledgeable worker was at least potentially threatening to the status and self-esteem of other representatives. Overall, however, the workers' seriousness, competence, and self-confidence usually have earned them respect.

Conditions favorable to the pet-food experiment

Listed below are eight factors which facilitated the success of the new plant.

1 The particular technology and manufacturing processes in this business provided significant room for human attitudes and motivation to affect cost; therefore, by more fully utilizing the human potential of employees, the organization was able to both enhance the quality of work life and reduce costs.

2 It was technically and economically feasible to eliminate some (but not all) of the routinized, inherently boring work and some (but not all) of the physically disagreeable tasks.

3 The system was introduced in a new plant. It is easier to change employees' deeply ingrained expectations about work and management in a new plant culture. Also, when the initial work force is hired at one time, teams can be formed without having to worry about cliques.

4 The physical isolation of the pet-food plant from other parts of the company facilitated the development of unique organizational patterns.

5 The small size of the work force made individual recognition and identification easy.

6 The absence of a labor union at the outset gave plant management greater freedom to experiment.

7 The technology called for and permitted communication among and between members of the work teams.

8 Pet foods are socially positive products, and the company has a good image; therefore, employees were able to form a positive attitude toward the product and the company.

Innovative Restructuring of Work

Richard E. Walton

In today's climate of worker malaise, many work organizations are exploring the idea of a basic restructuring of work to meet both the changing expectations of employees and to improve performance. It is increasingly apparent that employee alienation is not a "passing phenomenon," and that it is at the root of such critical workplace problems as high turnover, low productivity, poor morale, and sometimes even sabotage.

Recognizing that the costs of alienation are borne by both the workers and their employing organizations, a limited number of companies launched, in the 1960s, experimental work systems that were designed to attack both sides of the problem. These innovative systems strove to achieve a total, "systemic" restructuring of the way work is done.

This chapter assumes that parties wishing to take similar initiatives can benefit from the experiences of the early innovators. The first section argues that employee alienation is a basic, long-term, and mounting problem and hence warrants solutions of comparable form. The second, third, and fourth sections treat various aspects of a class of pilot efforts to solve the problem, analyzing their common features, how they were introduced, their results, their long-term viability, and their diffusion. These sections are based on a preliminary review of twelve pilot

---

Reprinted from Richard E. Walton, "Innovative Restructuring of Work" in THE WORKER ON THE JOB: Coping With Change, Jerome M. Rosow, editor, © 1974 by The American Assembly, Columbia University, pp. 145-176. Reprinted by permission of Prentice-Hall, Inc., Englewood Cliffs, New Jersey.

experiments in eleven different companies. Observations are based partly on field research in many of these firms under my supervision and partly on accounts developed by others as listed in a note at the end of this chapter.[1]

Experimental projects were included in the sample only if (a) they involved a relatively comprehensive restructuring and included operator-level personnel, (b) had existed more than two years, (c) were judged by their originators to have been initially effective, and (d) some satisfactory account of the project was available to me.

The sample includes more of the experiments which have received substantial publicity in recent years, although it does omit some notable firms which have innovated in this field. Of the eleven firms, four are in the United States, two in Canada, one in Great Britain, three in Norway, and one in Sweden.

COMPREHENSIVE REFORM

Systemic redesign of work systems involves the way tasks are packaged into jobs, the way workers relate to each other, the way performance is measured and rewards are made available, the way positions of authority and status symbols are structured, the way career paths are conceived. Moreover, because these types of changes in work organizations imply new employee skills and different organizational cultures, transitional programs must be established.

Before describing these features in some detail, let me provide some additional background regarding the sample on which my analysis is based. As I mentioned earlier, it includes twelve experiments in eleven different companies.

The United States Companies—An early innovator, Non Linear Systems, Inc., instituted changes in the early 1960s affecting the entire workforce of this small instrument firm. It is one of a few in the sample that explicitly abandoned the experiment and returned to conventional organization.

Donnelly Mirrors, Inc., is another early innovator that introduced incremental changes throughout the 1960s.

An experiment at Corning Glass was initiated in an assembly plant in Medfield, Massachusetts, in 1965. In the fourth United States firm, General Foods Corporation, detailed planning for change began in 1968; the experiment was initiated in a new pet food plant in Topeka, Kansas, in January, 1971.

[1]I have a current, albeit in some cases incomplete, understanding of the majority of situations. However, in a few cases the written accounts on which I must rely were prepared in the late 1960s or early 1970s.

The Canadian Companies—These are Alcan (aluminum) and Advanced Devices Center, a division of Northern Electric Company (subsequently renamed Microsystems International, Ltd. In 1964, a group of Alcan managers concentrated their innovative efforts on one of the fabrication plants in the works at Kingston, Ontario. At about the same time, Northern Electric was designing a radically different organization for a new facility which was occupied in January, 1966.

The European Companies—These form the remainder of the sample. Shell U.K. introduced change in several locations in the mid-1960s, including two of the plants in this sample—an established wax plant in 1966 and a new refinery at Teesport which came on stream in 1968.

Three projects were carried out in different industries under the Industrial Democracy Project, an action research program sponsored jointly by the Norwegian Federation of Employers and the Trades Union Council of Norway. Social scientists associated with the Work Research Institutes in Oslo also participated. The projects involved a department for assembling electrical panels at the Nobo-Hommelvik firm in Trondheim, a fertilizer plant at Norsk Hydro in Porsgrunn, and a department in the Hunsfos pulp and paper mill near Kristiansand. All three experiments were initiated in the mid-1960s.

A Swedish experiment in the truck assembly plant in Volvo is the final entry in the sample. This effort began in 1969.

All of the experimental units were manufacturing plants in the private sector. The manufacturing processes were continuous in slightly more than a third of the sample, assembly in another third; the rest were mixed, including batch processing.

More than half of the experimental units (plants or departments) employed between 100 and 500 employees; the other units had fewer than 100 employees. Their locations were about evenly divided between urban and rural small towns. More than half were unionized, but none of the plants in the United States sample involved unions.

Although all reports covered more than two years of operating experience, the actual periods varied from two years to over a decade. A third of the sample covered two or three years, a third involved periods from four to seven years, and the rest were eight or more years. In total they represented about 70 years of experience.

Finally, although each unit in the sample has a unique identity, as well as many differentiating characteristics, all units nevertheless have many features in common, not the least of which is the commitment to comprehensive workplace reform. In the remainder of this section, I shall discuss these features under three headings: (1) primary features, (2) secondary features, and (3) tertiary features.

Primary Features: Division of Labor

Central to all of the organization innovations considered here is the division of labor. In this regard, three design tendencies can be observed: work teams, whole tasks, and flexible assignment patterns. Let us examine each in turn.

Self-Managing Work Teams—These de-emphasize the idea of one-man/one-job in favor of groups that take collective responsibility for performing a set of tasks, as well as for some self-management.

The size of the team depends upon the nature of technological and social requirements. For example, many of the preassembly teams in Volvo were comprised of three to six workers. Typically, however, the teams contain from seven to fifteen operators, a size range large enough to include a natural set of interdependent tasks, yet small enough to allow face-to-face meetings for decision-making and coordination.

The reliance upon groups in these experiments is based partly on the belief in the power of face-to-face contact. This provides group members with social satisfaction and individual identification and develops goals, norms, and other capacities for self-management. Groups are also consistent with the idea of whole tasks and flexibility.

Whole Tasks—According to this concept, work that has been fractionated into simple operations is organized into more meaningful wholes that require more operator knowledge and skill. This may mean, for example, that the individual worker assembles whole units rather than merely adding one small part to the unit as it moves quickly through a work station. In continuous process departments, it could involve comprehension of and attention to a major segment of the process.

More significantly, tasks are made "whole" by incorporating functions that previously were performed by other service or control units. In every experiment the work teams took on substantially more inspection and quality control testing for their own work, often eliminating separate inspection departments or positions. The large majority of cases allowed operators to perform increasingly more of the maintenance on their own equipment. In batch processing departments, such as those in one plant in Alcan, operators set up their own machines. Custodial work—housekeeping the teams' own work area—is frequently included, especially where contamination of products is a critical problem.

In all cases, operators' work was designed to include planning as well as implementation, although the amount of planning responsibility varied widely. Scheduling of product runs and plant shutdowns for maintenance are examples of planning activities previously performed by supervisors or staff specialists that were assumed by work teams.

In the Nobo assembly department brief planning-reporting meetings involving all 30 employees were held each morning. Rotating coordinators took special responsibility for planning. Before the experiment, the planning concerns of workers were limited to their own individual activity on a daily basis. After six months, the workers' planning now effectively embraced the total group of 30 persons and covered a one-week time span. A year later, the planning was still on the group level but covered four weeks. After three years, worker planning had progressed to the point where it embraced five groups (the original experimental unit plus four others now brought into the system) and covered a three-month time span.

Members of teams often have additional enriching responsibilities. For example, the Volvo project included workers on consultation teams. Team members in the General Foods, Alcan, Shell, and other experimental units served in roles normally reserved for staff personnel or supervisors. These involved heading the plant safety committee, dealing with outside vendors, exchanging documentation with ship officers, and traveling to investigate or troubleshoot a customer's problem.

Whole tasks are consistent with various social-psychological ideas. For example, by integrating support functions into line groups, one eliminates many of the interfaces which tend to create intergroup friction. Also, the greater challenge in the resulting work is both motivating and confirming of self-worth.

Flexibility in Work Assignments—Flexibility is manifested in a variety of ways: (1) temporary reassignment from one position to another to cover for absences, (b) temporary redivision of work in order to handle a cluster of tasks at different manning levels, (c) progressive movement from one set of tasks to the next in order to master an increasingly larger segment of all work in a team and then in the larger experimental unit, (d) systematic rotation through a set of positions.

Flexibility has the obvious advantage of allowing effective use of available manpower, and it promotes individual skill development. The mutual learning helps reinforce coordination and teamwide planning activities. And the work team can usually decide how its members rotate through or learn a larger set of tasks.

Thus, the above three design tendencies constitute an internally consistent scheme for the division of labor. Teams make it possible to put together whole tasks. Identification with the team's "whole task" provides a rationale for learning all of the interrelated jobs. The flexibility options, in turn, provide an immediate decision function for cohesive, self-managing teams. Moreover, the personnel movement is more likely to produce psychological gains (e.g., variety, learning) that outweigh the costs (e.g., uncertainty) when it is self-managed by a group rather than controlled by a separate authority.

Secondary Features

The effectiveness of a division of labor with the above features depends upon the nature of the supervision and the design of the information and reward systems. The former should provide greater autonomy and enhance a worker's self-esteem and commitment to the work system. The latter should provide workers with the necessary tools to effectively assume greater responsibility and the appropriate rewards to achieve equity.

Supervision—Self-managing of teams requires that supervisors delegate many of their traditional functions of motivating, planning, and controlling. For example, in the pet food plant, the scope of decision-making delegated to the work team included: (a) coping with manufacturing problems that occur within or between the teams' areas, (b) temporarily redistributing tasks to cover absentees, (c) selecting members for plantwide task forces, (d) screening and selecting employees to replace departing operators, (e) counseling operators who did not meet team standards.

To effectively delegate such functions, supervisors must attend more to facilitating the processes and development of groups, longer-range planning, and external relations of the unit for which they are responsible. Sometimes they may do work previously done by their superiors or by staff units.

These changes imply basic shifts in the distribution of influence and expertise above the operator level. In the cases studied, the participative pattern of supervision was facilitated in two ways:

First, at least half of the cases involved significant trimming of supervision and staff. For example, several experiments omitted the first-line foreman and work groups reported directly to a general foreman. In one large works which included an experimental plant, the levels of supervision were reduced from seven to four.

Second, invariably supervisors were trained and coached to become more participative. Most managements reported career casualties among the supervisors who were expected to change—some could provide this type of leadership effectively, but others could not and lost their supervisory positions or were transferred.

Information System—While factory workers may be capable of assuming far more decision-making authority, they do not *auto-*

matically have either the analytical tools necessary to plan and control or the economic information needed to decide on the matters which fall within their enlarged sphere of influence.

Thus, the more sophisticated operations studied provided employees with economic and technical information typically given only to higher levels of supervision. In some cases, innovators discovered the necessity for new measurement procedures and operating feedback loops. The point is illustrated by the following conclusion reached during the design stage of the Hunsfos chemical pulp experiment:

> Firstly, considering the qualitative aspect of pulp production, we found that, among the output criteria most relevant to process control, only degree of digestion, brightness and tearing strength were measured systematically by the laboratory technicians. Cleanliness was judged subjectively from special test sheets, but factors such as pitch and homogeneity were too expensive or difficult for regular measuring. While there were no measurements on the quality of the input chips, information about pH value and percentage of sulphur dioxide in the acid were rarely based on statistical calculations. Because of the great variances observed in some of the quality measurements of individual batches, it was difficult to reveal long-term trends in the process control. The lack of feedback on this level reduced possibilities for continuous learning and control (Englestad).

Those who were restructuring the work in this pulp unit proceeded to clarify and define measurements of the quality and quantity of incoming and outgoing materials as well as quality control limits for each performance criterion. Gradually the measurements were instituted.

The experimentors used two general methods for exchanging current information: (1) regular meetings of operating teams or their representatives, supervision, and sometimes engineering personnel; and (2) an "information center" that posted all information about the state of the system.

Space factors also played an important role. Work team meetings and larger assemblies require space which can accommodate the right number of people and is free of noise and other interferences. The design of the Northern Electric system gave special attention to this aspect, probably because it focused heavily on professional and managerial personnel and a large variety of communication patterns was envisioned.

Reward System—The greatest diversity among the innovative systems studied was in the nature of the compensation system.

All of these systems involved an increase in worker duties and responsibilities. The worker used more of his faculties and accepted more responsibility for the performance of some unit of the organization. In such a situation, what is the quid pro quo for his extra investment? He is rewarded intrinsically through satisfaction from the work itself and the pride associated with higher status. He is also rewarded extrinsically—e.g., through increased pay and economic security. This subsection is primarily concerned with the latter. There are four important aspects of extrinsic rewards:

1. *The form of payment.* Most pay schemes in the systems studied involved a move from hourly wages to salary and made pay treatment for plant workers parallel to that previously afforded clericals, professionals, and other nonexempt office workers. Among other effects, this move tended to assure stable earnings over longer periods of time.

2. *The assurance of employment.* Because employment insecurity can undermine worker efficiency and commitment, many managements took steps to assure workers they would not be laid off for lack of work.

3. *The levels of individual compensation.* How these are determined is a critical issue. The following plan was employed in two of the experiments—the General Foods pet food and Norsk fertilizer plants—where employees manned a continuous processing technology.

Individual pay increases were geared to an employee mastering an increasing proportion of jobs, first in the team and then in the total plant. Increases also could be obtained on the basis of technical skill—e.g., electric maintenance in the General Foods plant—or increased theoretical knowledge—e.g., chemistry or measurement technique in the Norsk plant. Team members were, in effect, paid for learning more and more aspects of the total manufacturing system. Further, because there were no limits on the number of operators who could quality for higher pay brackets, employees were encouraged to teach each other.

This plan, of course, contrasts with the traditional wage schemes which feature large numbers of differentiated jobs and numerous job classifications, with pay increases based on progress up the job hierarchy. Who, then, decides when an employee has qualified for a higher pay bracket? In the General Foods plant, these decisions were made by the first-line supervisor, usually after consultation with team members. In the Norsk unit, the general foreman and shop steward made joint judgments.

On the one hand, this form of pay progression in the absence of a job hierarchy reinforces the personnel development required by a flexible division of labor. It also provides an incentive for one of the more meaningful forms of interpersonal cooperation, namely, teaching and learning.

On the other hand, the plan is not easy to administer. The evaluative implications of awarding different rates of pay can stir strong emotions in people who work so closely with each other. Are the judgments about job mastery appropriate? Does everyone have an equal opportunity to learn other jobs? Do judges depart from job mastery and other specified criteria and include additional considerations in making "promotions"? Such questions naturally arise.

Another consequence of this plan can be a rapid rate of pay increases, relative to rates of progression in conventional work systems. In the General Foods plant, for example, many employees were able to learn nearly all of the operations in the plant in two and a half years. They progressed from starting rate to near top rate in that time period—thereby earning a very substantial pay increase. While the obvious effect of such rapid progression is an increase in the overall level of worker compensation, many employees in the General Foods plant still did not perceive the plan as an adequate quid pro quo for the level of involvement and contributions they were making. In mid-1973, they were exploring group bonus schemes as vehicles for additional awards.

In the Corning plant, management's response to the "escalating climate of work involvement" was to change "toward a pay process more clearly based on merit (including appraisals for hourly and weekly salaried clerical and technical employees as well as for managerial and professional personnel)" (Beer and Huse). I do not know whether this scheme resulted in substantially higher average compensation levels or merely a distribution of pay more closely linked to differential performance.

4. *The level of overall compensation.* This probably becomes the most significant issue over the long term. (I emphasize "over the long term" for reasons to be treated later.) The quid pro quo involving higher compensation levels is related to various aspects of the work system itself.

The simplest relationship is illustrated by Non Linear Systems. In this experiment, management simply pegged wage rates higher to generally compensate for extra worker involvement. But other relationships are more complex:

Time in lieu of pay: "Compensation" may have to be taken in forms other than money. For example, wage rates for pre-assemblers and manning tables for certain pre-assembly units in the Volvo truck division were fixed in the union contract; therefore, employees were not able to earn additional compensation, but could consume the fruits of increased efficiency via the length and convenient timing of work breaks.

Higher pay for flexibility: The Shell, U.K. experiments were initiated while productivity bargaining between management and the union was being conducted. Management was seeking to exchange pay increases in return for a relaxation of restrictive work rules, and the types of flexible work patterns involved in the experiments were radical examples of what management was trying to negotiate with the union. Thus, the experiments not only provided both union and management with experience related to the bargaining but also were conducted in a context in which it was clear that the innovations would involve increased rates. And they eventually did.

Cost reduction formula: Compensation levels were linked to credible promises of future performance in one case. At Donnelly Mirrors, management related the size of the annual increase in base pay to (a) collective judgment about the feasible magnitude of cost reductions over the next year and (b) collective commitment to achieve the cost reductions that would pay for the increase. This innovative system also provided for group bonuses.

Cost and quality bonuses: Compensation levels were often linked to the work systems' actual results. In three Norwegian experiments, as with Donnelly Mirrors, group bonuses were a central feature. Two of these involved somewhat complex formulas incorporating significant factors which workers could influence. The fertilizer-factory bonus was based on production volume of acceptable quality, loss of nitrogen, total man-hours for production and service workers, and certain other cost factors. The chemical-pulp department bonus was paid on cleanliness, tearing strength, and degree of digestion and brightness.

The foregoing diversity in the provision for rewards reflects, I believe, a lack of consensus among the designers of the various work-system experiments. Many managers and psychologists tend to play down the matter of extrinsic rewards. Indeed, in the *short run* extra pay in the form of higher individual rates or group bonuses may not be necessary to elicit involvement. Short run, the intrinsic rewards involved in change, challenge, and personal development are sufficient. In the longer run, however, a sense of equity and sustained effort by workers requires a sharing of the extra fruits of productivity.

Tertiary Features

Another set of factors that help create appropriate attitudes and skills for systemic redesign of work involve symbols of status and trust and methods of training and recruitment.

Symbols of Status and Trust—Almost every experiment had features expressly designed to enhance the status of operators and to communicate trust in their exercise of self-control. Typically, there were no time clocks and workers were placed on salary and given the normal privileges of salaried employees.

Other devices were employed to minimize status differentials. For example, the General Foods pet food plant had an open parking lot, a single entrance for both office and plant, and a common decor throughout the reception area, offices, locker rooms, and cafeteria.

Training and Recruitment—The innovative systems studied also required greater technical and social skills than conventional systems. These were acquired by a combination of formal training, on-the-job learning, and recruitment.

During the period when the systems were being established, management provided considerable amounts of formal training. In a few cases, however, longer-term manning levels allowed 15 to 25 percent of employee time in continuing training activities.

Where the external labor market is favorable, a firm can tailor the work force to the innovative system, in terms of both a capacity for development of multiple skills and a receptivity to cooperative social patterns. Some companies in the sample did attempt to recruit workers with skills and interests that were consistent with the new system's requirements. In several cases, where new facilities were being manned, management selected from an unusually large pool and also provided applicants with information which encouraged appropriate self-selection. The General Foods pet food plant screened over 600 people to select about 70. And the Shell, U.K. refinery had 3000 men apply for the 156 jobs available. A recruiting advertisement for the Norsk experiment is illustrative of the ads used by General Foods and Shell:

> We need workers to take care of process and maintenance in the new fertilizer factory (process-workers, maintenance-workers, pipers, eventually instrument makers). The company is going to try to develop new kinds of cooperation to the benefit of employees as well as the company itself. Therefore we want to get into contact with employees who are interested to
> a) learn and develop themselves further through the work
> b) take responsibility
> c) become active members of a work-group
> d) participate in the training of others
> e) participate in developing jobs and ways of cooperation which create conditions for personal development through the work.
>
> It may be necessary to alter many of the usual norms within the organization, such as formal organization and contents of the different jobs. At the moment, it is considered probable that work-groups with optimal competence within maintenance and process-control will have to be formed (Bregard, et al.).

Although the Norsk innovators decided against getting "an elite group of men into the factory, but rather to find persons with qualifications close to the average in the company" (Bregard, et al.), the advertisement and the screening procedure probably did produce an above-average group, at least in terms of receptivity to the innovation.

Except for the foregoing three cases, there were relatively few attempts to preselect a work force with skills and attitudes especially appropriate for the innovative system. And at Northern Electric, where the new system subsequently gave way and returned to more conventional modes, the originators place part of the blame on their failure to give more attention to selecting employees for "fit."

The Need for Consistency

As indicated in the foregoing discussion, there is a strong need for consistency between primary features and support systems in the work environment. A change that is intended to embrace the major aspects of the work situation may in fact turn out *not* to be sufficiently comprehensive. In short, unless all aspects of an organization send similar signals, workers will sense the "ambivalence" and become frustrated in their attempts to respond to the innovation. The following inconsistencies are illustrative:

The appropriate leadership may not be ensured. The autonomy and self-regulatory aspects of work teams must be reinforced by the supervisory leadership pattern. This may take time, however, because supervisors with authoritarian personality traits will likely find it difficult to meet new role requirements.

The information system may be inadequate for the goal-setting and decision-making roles of work teams. Team members must understand business criteria, have competence in analytical techniques, and receive timely information about other units. The latitude to make decisions without the relevant skills and information creates an inconsistency that weakens the organization.

The reward system may not provide reinforcement for the behavior prescribed by other aspects of the innovative organization. For example, the work system may require a lot of mutual assistance among team members yet reward personal development rather than group performance. Another example is where greater skills and responsibilities enable workers to increase productivity but the reward system does not give them a share of the increased productivity.

To avoid these and other errors of inconsistency in innovative work systems, the primary, secondary, and tertiary features must be mutually supportive. Thus it is crucial to understand their interrelationships. The target of the innovation, of course, is enhanced quality of work life for employees and improved effectiveness for the organization.

INTRODUCTION OF THE INNOVATION

The impetus for the experiments took a variety of forms. In over half the cases, there was a philosophical commitment, albeit often tentative, to create more humane and effective work systems. In several instances, the experiment was fostered by a key manager's strong interest in the behavioral sciences. In many cases, persistent, sometimes chronic problems of turnover, productivity, and morale prompted innovation. Frequently, the impetus came from a combination of the above sources.

Method

A general pattern did emerge, however, in the way companies pursued organizational change. Typically, a limited experiment was conducted in a unit of the corporation. The experiment was conceived as a pilot project from which the larger organization could learn. (Two exceptions to this pattern were the relatively small, owner-managed firms—Non Linear Systems and Donnelly Mirrors.)

At Alcan and Volvo the experiments were introduced into particular units of a large existing complex. Subsequent changes were instituted incrementally over many years. Similarly, one project at Shell, U.K. involved an established unit.

Just as often, however, new concepts were introduced as part of a new unit at a new site. Cases in point are the Corning instrument plant, the General Foods pet food plant, one of the Norsk plants, and the Shell refinery. A slight variation on this pattern occurred at the Northern Electric project, where the introduction of the innovative system coincided with a move from Montreal to Ottawa and a doubling of the work force.

Before discussing the early results of these pilot programs, I should note that a number of conditions are especially favorable to their successful implementation. Based on the case studies, I have isolated seven conditions which seem especially important. These are listed immediately below:

Seven Conditions Favorable to Pilot Project Implementation

1. Typically, small towns provide a community context and a work force that is more amenable to the innovation. Half of the experiments were implemented in this type location.
2. Smaller work forces make individual recognition and identification easier. Half of the initial experiments involved fewer than one hundred employees.
3. It is easier to change employees' deeply ingrained expectations about work and management in a new plant culture. About half of the experiments were in situations of substantial "newness."

4. Geographic separation of the experimental unit from other parts of the firm facilitates the development of a unique plant culture. Advantageous geographic separation appeared to be a factor for the pet food plant, the refinery, and the assembly plants of Nobo and Corning.
5. The use of outside consultants as change agents provides objectivity and know-how to the experiments. The majority of the firms had a pattern of using outside consultation in organization development, knew how to use this type of assistance, and were not subject to criticism.
6. The long lead times that are implicit in start-ups allow large blocks of time for training and acculturation. This was a significant factor in several cases.
7. Where there is no union, or where union-management relations are positive, it is much easier to introduce the type of work systems studied. The unionized seven plants had positive union-management relations when the experiment was undertaken. Here, the parties typically agreed to a "sheltered" experiment, in which the normal contract provisions and practices were relaxed for a limited time period, and the changes would not set precedents for other units and that the experimental unit would return to its earlier pattern in absence of mutual consent.

Early Results

By design the sample included only experiments which yielded at least initially positive results. In this section, I shall review the types of results claimed for these experiments, even though, with one exception, I cannot vouch for the validity of the claims made by the participants or researchers. The exception is the General Foods pet food plant where I was closely involved as a consultant.

Two cases—Nobo and General Foods—illustrate a middle ground between the more modest and more ambitious claims of early results:

1. After one year, a follow-up study at Nobo showed that (a) the group system had been transferred from the original group of 30 to four additional groups totaling over 100 employees; (b) productivity went up 20 percent; (c) quality control and other service activities were satisfactorily decentralized to the groups; (d) the time perspective of workers increased from three hours to three months; and (e) only a small minority preferred the old system to the new group system, where respondents were referred specifically to variation in work role, job learning, participation in decisions, relations with work mates, relations to company, and breadth of responsibility.

2. The pet food plant had a manning level of fewer than 70 people, rather than the 110 estimate based on industrial engineering standards. This difference resulted from the team concept and the integration of support activities into team responsibilities. Further, after 18 months, the new plant's fixed overhead rate was 33 percent lower than in the old plant. Reductions in variable manufacturing costs (e.g., 92 percent

fewer quality rejects and an absenteeism rate 9 percent below the industry norm) resulted in annual savings of $600,000. The safety record was one of the best in the company and the turnover was far below average. While new equipment was responsible for some of these results, more than one half of them derived from the innovative human organization. Operators, team leaders, and managers alike had become more involved in their work and also had derived high satisfaction from it. For example, when asked what work was like in the plant and how it differed from other places they had worked, employees typically replied: "I never get bored." "I can make my own decisions." "People will help you; even the operations manager will pitch in to help you clean up a mess—he doesn't act like he is better than you are."

Quality of Work Life—All experiments reported early improvements in the quality of work life, although the degree of positive employee responses in the early period was quite varied, as were the patterns of these responses over subsequent time periods. Individuals differ in their preferences for variety versus routinization and stability of job-related duties. These differences have been found in survey studies and have been reported in some of the innovations reviewed here.

One administrator in the Northern Electric Company tried to summarize the difficulties inherent in implementing the new approach:

> A lot of people felt it would be automatic, just by changing the structure. The weakness was the assumption that people would be highly motivated in this kind of environment and they're not. There's a threshold value of personal maturity or outlook, and below it people are more effective in a hierarchical, not an open system. The . . . system doesn't take into consideration the differences in basic behavior of people. Some will accept the freedom and thrive in team organizations. But others are just not responsible or self-disciplined enough to make this work. These differences are not divided according to discipline or education. It's a function of personality and it's in all areas. I'd say that there are about 25% who truly respond properly in the participative sense. The other 75% don't (Gabarro and Lorsch).

These individual differences raise the question: How many persons do gain significantly from the changes involved in the innovative organization? The answer obviously varies from one group of employees to the next and therefore must be assessed on a case-by-case basis.

For example, in sharp contrast to the above Northern Electric estimate of a 25-percent positive response, I judge that approximately 80 percent of the General Foods workers experienced relatively large gains in the quality of their working life relative to their work histories. The balance of the experiments in the sample would fall somewhere between the Northern Electric and General Foods statistics during their initial periods.

Organizational Performance—Early gains in organizational performance were almost uniformly reported. For example:

1. Higher production efficiencies were reported in at least six cases; these were derived from less wastage of materials, less down-time, or more efficient methods.
2. Quality improvements were significant in five cases.
3. A reduction in overhead was common—for example, due to a leaner supervisory and staff structure and less paperwork.
4. In several cases, the more rapid development of skills produced promotables at a more rapid rate, increasing the number of operators who were promoted to foremen outside their assigned department.
5. Turnover and absentee rates were usually reduced.

An excerpt from a report on the Shell, U.K. wax department helps to illustrate the interdependence of worker satisfaction and organizational performance when the innovative work system is applied to continuous processing technology:

> The finishing unit capacity was increased by 40% over the 1965 figure again primarily due to technical improvements—some of which were suggested at the Department Meetings.
>
> . . . the understanding and knowledge which the operators had of the changes made played an important part in maximizing the gains made possible by the changes.
>
> The most important increase in output was in the second and third units. . . . Important because this is the limiting section of the plant and output was increased 100%—but most important because it is largely manually controlled. This part of the plant is extremely demanding because it has sixteen operating variables which must be closely controlled. Of course controls themselves are instrumented but the inter-action of the controls cannot be because of the nature of the process. To optimize such a process required knowledge of the operation, manual skill and constant attention to the job. As the output increases so does the demand for attention to detail with the need for frequent minute adjustments to the plant. This is the type of job which can only be successfully accomplished by highly motivated operators—and the output increase indicated is a measure of the success achieved. There is no doubt that provision of laboratory testing facilities on the plant helped motivation—because the operators were able to get immediate feedback on the results (which) is satisfying psychologically. This point was made to me many times by the operators concerned—but they felt they knew how well they were doing, and this encouraged them to go on. The whole unit appeared to be under their control—and that is just what we were aiming for (Burden).

PROBLEMS OF SURVIVAL AND GROWTH

Thus far, we have been exploring *initial* design features, *introduction* of the innovation, and *early* results. What has happened over a longer time frame? As I pointed out earlier most of these experiments were pilot projects in particular units of larger organizations. Originators of a project expect that if it is initially successful, the innovation will continue to evolve toward its original ideals, and be emulated elsewhere. Thus we distinguish three aspects of the development of organizational innovations: (1) origination, (2) continuation, and (3) diffusion.

The rationale for continued viability and evolution is that the original change will develop a plant culture with the values which underlie the innovative organization. Further, it is expected that the results will reinforce both the participants' involvement and their superiors' support.

Diffusion is projected because it is assumed that an organization pattern which works better than its predecessor will be recommended by superiors and emulated by peers. In practice these two projected tendencies of continuation and diffusion are often complicated or nullified by a host of other dynamics. Although I have not completed the process of establishing measures of whether, and to what extent, the innovations in my sample have reverted to more traditional patterns after an initial period of successful change, the evidence permits some rough summary observations.

At least three plants have returned to conventional patterns. Serious efforts are being made to revive one of these. Many others have regressed somewhat after a few years of successful evolution toward the ideals underlying the innovation. The work situations still remain, however, significantly unconventional. Several other innovative plants, as of this writing, are still successful and evolving in the direction that they were launched.

What has tended to undermine those innovations which have terminated or regressed? In the cases of continued success, what factors and dynamics, if any, have threatened the system? What can be done to minimize these threats?

Lack of Hierarchical Support

In many of the cases studied, a higher official had in effect "held an umbrella over the experimental unit," protecting it from premature evaluation and absorbing some of the risks involved in the venture. When the higher executive was replaced by one who was not sympathetic, the personnel in the unit felt increased career risks.

In some cases, innovation raised expectations of employees, who were subsequently disappointed because management failed to follow through. The Norsk case is illustrative:

The Norsk experiment after four years revealed mixed results. Productivity had gradually increased and down-time was cut by more than half. Worker satisfaction was still relatively high. The majority of workers had mastered all blue-collar skills in the factory. Yet, despite the increase in worker competence, management had been unwilling to reduce supervision and materially increase the workers' influence in critical decisions. According to Gulowson, the failure to delegate more supervisory tasks to workers, together with a decline in the incentive of learning new operator tasks (as mastery levels climbed), had made monotony a problem once again. Gulowson concluded:

> The experiment has demonstrated the conservatism of large organizations. To the extent that the experimental area has been dependent upon changes in the environment, these changes have only seldom been made. In terms of total system behavior, the environment has forced the experimental system almost back to where it started.

Thus, to promote the viability of the innovation, higher management must sustain philosophical support over a number of years and must be prepared to make further organization changes as they are indicated.

Loss in Internal Leadership and Skills

Turnover in leadership within a unit has created problems. The top position in one experimental unit was refilled four times in six years, during which the innovative system almost died out. Subsequent leaders did not take decisive actions to recreate a more conventional form, but acted in ways inconsistent with the spirit of the innovation, and the result was the same.

In several instances, where the experiment was introduced during a plant start-up, training of operators was begun well in advance, and learning through errors was tolerated during the early phases of the start-up. Thus, the bank of necessary skills was built.

But the requisite bank of skills cannot be maintained if the turnover rate exceeds a certain threshold level. This threshold rate appeared to be less than 10 percent in one case. After the Teesport experiment was underway, management found it necessary to reduce the range of work flexibility among team members and put back a level of supervision. Hill's account of this partial retreat in the design mentioned the relatively high turnover of 10 percent along with technical difficulties and additional tasks. Hill attributes the 10 percent rate to the existence of unusual opportunities for overseas assignments.

The loss of skills through turnover not only can promote a scaling down of an innovative project but also can arrest the development of an overall change program, such as that launched by Shell, U.K. in 1965.

During 1967-68, many changes in job assignments occurred, reducing the skilled resources available to facilitate the changes initiated. For example, of the eleven senior managers who participated in the original planning conference, only six remained. Also, two principal consultants departed for overseas. Although some changes were unavoidable retirements, others resulted from career changes.

Hill concluded that this severe dispersal of resources "undoubtedly hindered a fuller realization of the potential developments at that time."

I believe that such turnover among the leadership of innovative units, the expert staff, and the consulting resources is a natural tendency. For example, I would expect the leaders of organizational innovations to be relatively able and secure as persons. Thus, if the experiment should show signs of success, these leaders would become even more visible, and new career opportunities would present themselves.

These innovative organizations appear to require greater stability of personnel than the conventional organizations they replace. Under normal circumstances the attractive aspects of the work system will tend to produce low turnover rates among workers, thereby ensuring this particular condition for its own success. If higher turnover rates are nevertheless encouraged by exogenous factors, management must take extra steps to stabilize the work force. The problem of turnover of key managers, staff personnel, and consultants is a different matter—where success undermines itself. *Therefore, it is desirable to secure longer-term commitments from key personnel at the outset.*

Stress and Crises

New demands may also tax the system's ability to perform and survive, producing a return to more conventional patterns. The cases illustrate two types of demands: technical problems and competitive pressures.

Technical Problems—As I mentioned above, the necessity to put back a level of supervision and reduce the range of job flexibility in the Teesport refinery was attributed not only to the turnover but also to (a) the expansion of the work performed in the refinery and (b) technical problems. According to Hill, the technical problems "tended to prevent the establishment of steady state operation . . . and induced a certain amount of unexpected stress in the social system." The social stress placed a premium on more predictability; and certainty was sought through less movement of personnel, more specialization among workers, and closer supervision.

Competitive Pressures and the Survival Syndrome—"Survival" patterns developed in several firms sometime after innovative work systems were launched.

Two of these companies came under new, severe, and long-term competitive pressures that resulted in new initiatives and influence patterns emanating from the top. Higher management began emphasizing cost reduction and near-term results, insisting upon discipline and compliance with their programs, and in general providing an inhospitable environment for the innovative work system.

Authoritarian decisions and "do it" commands tended to erase the premise that a subordinate could freely challenge superiors in unguarded dialogue. Politically based influence techniques undermined the premise that a person's influence would be a function of his expertise and information. And, as cliques formed to exercise influence, interpersonal relationships were corrupted, trust was eroded, and the sense of "community" began to deteriorate.

In a third case, Non Linear Systems, the organization returned to a conventional form when revenues dropped rapidly in 1970 and 1971—from $6 million in 1965 to $3.5 million in 1971. Significantly, this slump was a part of a general downturn in the aerospace and many related high-technology industries. There are differences of opinion about which, if any, of the innovations instituted many years earlier contributed to NLS' downturn and which ameliorated the decline. The president, Andrew Kay, blames the experiments for the fact that he delegated so thoroughly and lost touch with the operational aspects of the business. Yet rebuttals of some former executives and consultants indicate that certain developing realities in the business were brought to Kay's attention and that he made a personal choice to believe what he wanted to believe, namely his "dream."

Fortunately, for the present analysis, one need not determine whether the Non Linear Systems innovations were a hindrance or a help in coping with the economic downturn. Rather, one is satisfied to conclude that the business crisis contributed to the decision to abandon the experiment. (This is not to deny that other factors, including the personality of the president, undoubtedly must be considered in explaining both the origination and the subsequent termination of the innovation.)

Participants in an innovative system and higher management must be alert to the regressive tendencies that accompany stress and crises.

If, for example, the situation appropriately requires more direction from the top and extra measures to avoid mistakes, these needs can be discussed at all levels and the demoralizing effect of the changes can be minimized, thereby preserving the widespread commitment to return to precrisis patterns after the crisis subsides.

Tensions in External Relations

Parties external to the experimental unit often become increasingly concerned about how the unit functions, apart from how well it performs. These parties include superiors, peer departments, staff departments, labor unions, customers, and vendors. Their impact ranges from the capability to declare an end to the experiment to the ability to force demoralizing compromises. Nevertheless, their preoccupations are similar: "How much may the innovative unit be allowed to deviate from general practices?" Thus the dilemma for an innovative system is between maintaining internal integrity and external consistencies.

Equity Issues—Unresolved equity questions can result in damaging pressures from outside parties. Consider, for example, the Teesport refinery. In establishing the refinery experiment in 1966, management persuaded the union to negotiate a local contract that was separate and completely different from the existing national contract. Although both parties recognized that changes might be needed in the future, the contract allowed the freedom necessary to conduct the experiment. With this latitude, the parties agreed to higher pay and other favorable terms to match the additional skills developed by employees and the extra work flexibility and responsibility.

Later (1968-70), management and the union engaged in productivity bargaining at other work sites. The bargaining in other refineries

> tended to bring their terms and conditions of employment—such as annual salary and staff status—closer to those of Teesport without, however, approaching the level of job flexibility and responsibility achieved at Teesport. The effect of this partial closing of the gap has been to create pressure to close it further *by moving the Teesport job structuring back towards the conventional norm* (Hill; emphasis mine).

Consider the implications of the last statement in the preceding paragraph. It suggests that the equity concept is so strong that even though employees may be intrinsically rewarded by taking on higher responsibility and making high contributions, their extrinsic reward must also be in line with their relatively high work inputs. If this is not the case, their sense of injustice will cause them to scale down their level of involvement.

Pressures for Uniformity—In almost every case, the experimental unit came under strong pressure from peer and staff units to conform to company policies and conventional norms. The pressure often mounted after an initial period of grace, when it appeared that the innovations were not failing of their own weight and might become permanent.

These strong conformity pressures are probably inevitable, but I have observed a particular tendency on the part of members of some experimental units which exacerbates the problem. It is sometimes referred to as an "evangelical syndrome"—a "holier than thou" stance toward outsiders who are as yet still "conventional" in their approach to work organization. This "superiority complex" causes bosses and staff units to pressure the unit back into line and peer units to reject similar change for themselves.

Collective Bargaining Dynamics—In one case, union relations contributed to the "undoing" of the innovation. Many features of the new work system had been introduced during a period when, according to management, the union leadership had been stable and politically secure. The philosophy of the earlier leaders was illustrated by their pride in the absence of formal grievances. Later, however, the union became more legalistic—and tended to write up all complaints. Management believed that the union officials became too insecure to sort out legitimate grievances from those where an employee was simply trying to take advantage of a loophole or the informal understandings which had developed on the work floor. There were signs of a vicious cycle:

A foreman feels exploited if he observes an employee abuse the informality (of no time clock, for example) by coming late and then making a claim for a short period of overtime on the other end of the shift. If the worker gets away with the claim, then more co-workers will say, "O.K., if that's the name of the game, I'll do it too." If the foreman cracks down, as he did sometimes, then a hard-nosed management is seen as turning away from the norm of informal problem-solving which characterized the innovative plant society.

Some of the above tensions in the external relations are avoidable. For example, the sponsors can avoid an "evangelical" effort to sell the innovative work system to sister units. However, to some extent these tensions are inevitable. Therefore, the sponsors must negotiate resolutely to preserve the integrity of the system.

Costs Versus Benefits for Individual Participants

A premise of the new work systems is that participants will gain when they accept greater responsibility and more complicated human interdependencies on the one hand and exercise more influence, receive social support, and enjoy personal growth on the other hand. The former typically have psychological "costs" associated with them; the latter typically yield psychological "benefits." Thus, one can speak of a more or less favorable or unfavorable ratio of (psychological) costs and benefits associated with the work organization.

Earlier, in the discussion of "results," I acknowledged that individual workers differed in their preferences for these potential benefits and costs associated with innovation. Moreover, the viability of several systems was threatened by such individual differences.

Designers of innovative work systems face a dilemma. If an innovative system does not accommodate those who prefer much less task responsibility, variety, and human interdependence than is idealized, and if these persons are selected out, then there is a tendency for the others to regard the system as a special case; this will limit its diffusion, confine its constituency, and isolate it. If, however, there is a tendency to press toward the least common denominator in terms of employee skills and readiness to accept responsibilities, then job design will tend back toward a conventional organization.

My recommendations would be (a) launch innovative work systems where the work force is generally favorable, (b) provide candidates with information which allows some appropriate self-selection, and then (c) cope with the minority who do not immediately "buy-in" by providing some diversity in work demands.

The problem associated with psychological costs and benefits may be even more complicated than this. The ratio may shift over time—with critical implications for the viability of the new unit.

I assume that the levels of human energy available for the organizational effort are related to a favorable exchange of psychological costs and benefits, and that the terms of these exchanges are altered as an innovation progresses. The following scenario presents some hypotheses consistent with my preliminary observations.

How is the human energy elicited for the initial experimental stages? The extra effort required for the learning, planning, and persuasion activities probably derive from the desire to create; prove it can be done; collaborate with others; get recognition; and learn and develop new skills. But this investment of energy does not pay off immediately. There is an early period of deferred gratification, while some suffer set-backs and others are taking a wait-and-see stance.

Assuming the innovation begins to "take" and produces encouraging results, there follows a period during which participants are highly interested in exercising more influence and accepting the responsibility that goes with it. During this period, participants are more ready to rise to the challenges of tasks which fully tax their capabilities, the demands of personal growth, and the trials of living with the uncertainties involved. They spontaneously devote the high level of energies all of this requires. There is also a keen sense of enjoyment in the personal flexibility allowed by fluid relations and low structure.

Participation in this type of innovation frequently is a vehicle for a major episode of personal development, creating the *commitment* source of energy. Yet, while desire for personal growth never ends, high levels of commitment to the growth process cannot be sustained indefinitely. Thus I suspect that the intense period of growth for the individual is of a shorter duration than the period over which the *organizational* energy requirements remain high.

Let's explore further the idea of a temporary period of high psychological benefit for participants in the innovation. I believe it is marked by a high visibility and attention from others because of its newness and novelty, a high stimulation from behavioral science consultants, and a high rate of demonstration of new competencies and mastery. If so, there must be a letdown when outside attention declines and the consultants leave, when one has learned about the related tasks, and when one is sharpening existing skills rather than acquiring new ones. Furthermore, during the period of high involvement, participants often limit some other life relationship—family, friends, hobbies, recreations. At some point, they want to rebalance their on-the-job and off-the-job commitments. And this desire is likely to sap the personal energy source.

Still another tendency can be hypothesized: at some point, the desire of the originators to promote the diffusion of the innovation may begin to compete with their desire to focus on further evolution of the original innovation. The diffusion task offers them the opportunities to develop new skills and to gain additional professional visibility.

The dynamics just hypothesized, if valid, involve subtle management issues. *It is desirable to take advantage of the extra energy elicited by the newness experienced by individual participants. Thought should be given also to ways of ensuring that a high level of energy will be sustained until the viability of the work system is secure, perhaps by infusing new participants.*

Isolation—Failure to Diffuse

Failure to diffuse changes made in the innovative unit to other parts of the organization can hurt the original effort. Without diffusion, the managers and supervisors in the new unit will have developed skills and experience that are perceived by others and by themselves to have only local relevance. Moreover, their desires for upward mobility in the company will conflict with their commitment to the ideals of the innovative unit. In short, either their enthusiasm will flag or their tendency to isolate the innovative system will increase.

In about half the cases studied, there *was* diffusion of the innovation from the pilot project to other units in the same firm. The significance of innovations in these "cousin" units has not yet been ascertained. Nevertheless, I can offer a few ideas about how this diffusion tends to occur and what tends to inhibit it:

Transfer of key personnel from the original innovation can promote diffusion. In several cases, a series of moves placed the leaders of an earlier experiment in high positions where they promoted diffusion into a number of additional units.

Visits by other interested groups to an experimental unit often result in the diffusion of innovative ideas. All of the experiments studied were visited, visited, and revisited.

Evangelism may be self-defeating and hierarchical support may decline, undermining diffusion efforts.

Rivalry may exist among organizational peers. In a number of cases, personnel involved in the experiment noted the very many visitors from outside the company and the few, if any, from other units within their own company. Unfortunately, rivalry is sometimes acute even among innovators in different units of the same firm. They share a common philosophy but work hard to differentiate their approaches. In several large firms this rivalry frustrated the development of an internal collegial network which could serve as a forum for exchanging ideas, identifying barriers to innovation, and exercising collective influence.

An innovation, successfully established in a plant unit, must be diffused or it may die.

CONCLUSION

Summary

Analysis of changing expectations of employees shows why we need major reform and innovation in the workplace. Employees increasingly want their work to be characterized by challenge, mutual influence patterns, dignity, positive social relevance, balanced attention to emotionality and rationality, and cooperative social patterns. In order to substantially increase these ingredients, the work situation must undergo comprehensive change. Piecemeal reforms, such as job enrichment, management by objectives, and sensitivity training are inadequate.

The organization redesign should be systemic. First, redesign must focus on the division of labor, involving, for example, the formation of self-managing work teams, recreation of whole tasks by reversing the trend toward fractionation of work, and an increase in the flexibility in work assignments by a variety of means.

Second, the redesign must embrace supporting elements, such as a trimming of supervision and more delegation of authority. Also, the information and reward schemes must be tailored to facilitate the delegation of decision making and to reinforce team work.

Third, other elements in the work situation must enhance the status of workers and communicate trust in their exercise of self-control—e.g., salaried payroll and no time clock. Similarly, recruitment and/or training are required to ensure the necessary skills.

Obviously, the revisions in these many elements must be coordinated and must result in a new, internally consistent whole.

The impetus for work restructuring experiments of this kind comes from prior philosophical commitment, an interest in the behavioral sciences, and compelling personnel or productivity problems.

A number of conditions are favorable to the introduction of such experiments: new plants with small, nonunionized work forces, located in rural communities geographically separate from other parts of the firm. None of these are necessary conditions, but each facilitates the rapid introduction of the innovative work system.

By design of my sample, the experiments reviewed in this study reportedly produced positive results in the first year or two of their existence—in terms of both quality of work life and productivity indexes.

However, several of the experimental units suffered set-backs after an initially successful introduction. A number of factors can threaten termination or create regression in these innovations: a lack of internal consistency in the original design; loss of hierarchical support; loss in internal leadership and skills; heightened stress and crisis; tensions with various parties external to the unit; an unfavorable ratio of psychological costs to benefits for individual participants; and isolation resulting from a failure to diffuse. With foreplanning, sponsors and leaders of innovative work systems can minimize the potential threats listed above.

Needed: Additional Experience and Knowledge

Managers, union officials, and workers themselves should be encouraged to join in efforts to redesign work organizations. In this effort, four factors require special attention.

First, we have relatively little experience with comprehensive work restructuring in unionized situations in the United States. Unions have tended to be, for a variety of reasons, suspicious of these schemes. They tend to see them as a threat to union existence. I do not. There are two core functions of collective bargaining that are not in any way eliminated by systemic redesign of work: (1) providing appeal mechanisms that ensure due process and (2) bargaining over factor shares. Hopefully, a number of United States unions will enter into "sheltered" experiments in the next few years so that they—and all of us—can further judge how innovation serves employee interests and affects labor as an institution.

Second, compensation is the least understood element of these new work systems. To what extent and in what form can workers be provided with a quid pro quo for the qualitatively different and greater contributions they make? How can one acknowledge individual differences and reinforce group level cooperation? Effective linkage of economic rewards with new work schemes reinforces the results.

Third, we do not know how to handle the fact of individual differences in designing new work schemes. Should entire plants have a uniformly challenging culture and the appropriate self-selection made at the plant level? Should smaller work units within a plant be varied, so a worker can find a unit with demands that fit his personal preferences? Or should each work group contain assignments that represent the full range of challenge, allowing for diversity within work teams? These issues deserve more attention.

Fourth, my preliminary research has confirmed what I suspected when I began my field investigation several months ago—namely, that even after successful introduction and early signs of effectiveness, a host of factors threaten the continued viability of the redesigned units and frustrate efforts to extend innovation to other units.

NOTE: The author acknowledges the research assistance of Richard Harmer and the support of a grant from The Ford Foundation; that the first section of this chapter reiterates an analysis in his article, "How to Counter Alienation in the Plant," *Harvard Business Review* (November-December, 1972), which also contains a detailed analysis of one of the plants mentioned here; and that other observations were based on the following accounts developed by others describing experiments in several companies reviewed in this chapter:

Bregard, A., Gulowson, O. Haug, Hangen, F., Solstad, E., Thorsrud, E., and Tysland, T., "Norsk Hydro: Experiment in the Fertilizer Factories" (Unpublished report, Oslo, Norway, January 1968, 23 pp.); Beer, Michael, and Huse, Edgar F., "A Systems Approach to Organization Development," *The Journal of*

Applied Behavioral Science (1973, 8 [1], pp. 70-101)—Corning Glass; Burden, Derek W. E., "A Participative Approach to Management: Shell, U.K. Ltd.—Microwax Department" (Unpublished report, April, 1970, 17 pp.); *Business Week*, "Where Being Nice to Workers Didn't Work" (January 20, 1973, pp. 99-100)—Non Linear Systems, Inc.; Engelstad, Per H., "Socio-Technical Approach to Problems of Process Control," in *Design of Jobs*, Louis E. Davis and James C. Taylor, eds. (Middlesex, England: Penguin Books Ltd., 1972)—Hunfos; Faux, Victor, and Greiner, Larry, "Donnelly Mirrors, Inc." (Copyrighted by the President and Fellows of Harvard College, 1973, 32 pp.); Gabarro, John, and Lorsch, Jay W., "Northern Electric Company (A), (B), (C), (D), and (E)," (Intercollegiate Case Clearing House, Harvard Business School, 1968, 14 pp.); Gulowson, Jon, "Norsk Hydro" (Unpublished report issued in April, 1972, by Work Research Institutes, Oslo, Norway); Hill, Paul, *Towards a New Philosophy of Management* (Tonbridge, Kent, England: Tonbridge Printers Ltd., 1971)—Shell, U.K.-Teesport; Thorsrud, Einar, "Democratization of Work Organizations; Some Concrete Ways of Restructuring the Work Place" (Working paper, Work Research Institutes, Oslo, Norway, June, 1972, pp. 6-12)—Nobo-Hommelvik.

Participation: Various Ways of Involving People in Their Work and Work Organization

N. V. Philips Europe

INTRODUCTION

In 1974 the Committee on Participation (COP) invited Philips' managers in European countries to complete a questionnaire about the spread of work structuring and work consultation in their organizations. Work structuring was defined by the Committee as "carrying out changes in tasks and the distribution of tasks in order to enlarge the autonomy and influence of all those involved".

The definition given for work consultation was: "the setting-up of consultation groups in order to widen the individual's participation in decision-making in the work situation".

Many replies were received. These showed that efforts are being made at various Philips' places in Europe to develop new forms of co-operation. However, the answers were not confined to the activities defined by the Committee as job structuring and work consultation; they also related to other aspects of company activity. One point common to them all, however, was that they could be regarded as attempts to increase the interest of those concerned in their work and in their work situation.

Excerpted from Chapters 1 and 3 of *Participation: Various Ways of Involving People in Their Work and Work Organization*, compiled by the Committee on Participation (COP) set up by the European Conference on Personnel Management (ECOP) to collect information about and stimulate new methods of involving Philips' employees more closely in their work. Used with permission of the publisher, N. V. Philips' Gloeilampenfabrieken, Eindhoven, The Netherlands.

The committee considered that as such they could be regarded as forms of participation.

In the present booklet we propose to give an impression of what is going on in Philips in Europe in this field.

Obviously the multiplicity of activities taking place in Europe made it impossible for the Committee to describe them all. A more reasonable idea seemed to be to make a selection of cases which—without laying any claim to completeness—give as accurate as possible a picture of the diversity of participation.

The selection of the cases was based on the greatest possible diversity of countries and technologies. The examples chosen are not, however, intended as *ideal* projects or as being *representative* of a particular country or technology. The main reason for aiming at the greatest possible diversity was to make it easier for others to recognise situations and problems similar to their own and thus to stimulate them to similar activities. The Committee therefore hopes that this booklet will give readers ideas for activities at various levels and in various places. They hope that it will encourage readers to learn from the experiences of others by reading more detailed accounts or reports as well as by establishing contacts with people active in this area.

Finally, the Committee hopes especially that this booklet will lead to intensive mutual discussion and to concrete efforts to involve employees more closely in their work and their work organization.

The Committee wishes to express their appreciation to all those who, by completing the questionnaire, made publication of this booklet possible.

CHAPTER 1

Some Considerations with Regard to Participation

The significance of the employer's job content, a reasonable degree of independence in doing that job, and his influence on his work situation have been underestimated for a long time.

The importance of these matters for an employer's attitude to his job and the working of the company has not been fully understood.

Efforts to improve material prosperity have meant that little attention has been paid to these aspects until recently. The employee was usually represented as someone who was only interested in his wages and shorter working hours. Work was often organized in such a way that a minimum of room was left for initiative and independence and little consultation and communication were required. It was scarcely realised that a work situation of this kind could create employees who were in fact interested principally in their wages and free time.

Combined efforts to increase material prosperity obviously yielded remarkable results, enabling the industrialised countries to achieve rapid growth of their prosperity for several decades on end. But that growth in prosperity was not without great tensions. Increased absenteeism, complaints about mental stress and a high turnover rate in certain employee categories were symptoms of it.

As prosperity grew—and with it the level of education and a critical attitude—the need for well-being also increased.

In this connection we need only think of needs on the personal plane: work that makes sense, room for individual freedom, opportunity for development and involvement.

The response which the Committee had to their questionnaire showed that these needs were being increasingly recognised in the various national Philips organizations in Europe and efforts were being made to satisfy these growing needs by making changes in: the work, the work organization, systems and procedures and the ways changes are brought about.

The Committee has put these activities under the heading *"Participation"* which is defined as meaning: *the creation or enlargement of the possibility for employees to influence their work and work situation*.

Participation in this sense may be seen as an essential element in all kinds of processes of change which are being brought about to enable both the organization and the individual to subsist and to acquire long-term viability.

A study of both internal and external investigations shows that there is hardly an experiment concerned with the process of change which fails to demonstrate that satisfaction in work is enhanced or that other consequences generally acknowledged to be beneficial are produced by a genuine increase in the employee's possibilities for participating in his work and work situation. When the need to increase the employee's participation is discussed, it is chiefly the factory personnel category that is thought of and then mainly the unskilled and semiskilled among them. That is understandable because in many cases it is particularly this group which is most restricted as far as having a say in their work situation is concerned.

With regard to *the need* for more participation, however, it is felt by the entire personnel and attempts to achieve greater harmonization between human expectations and aspirations on the one hand, and the company's possibilities on the other hand, must include everyone who has a stake in the company.

AMIENS, FRANCE*

Semi-Autonomous Groups in a Washing-Machine Assembly

Situation

The Amiens factory produces washing-machines. The production process is spread over four departments: the mechanical department, the surface-treatment department (enamelling and spray-painting), the electrical department and the assembly department.

The project described here concerns three assembly lines on which three different new models are assembled.

The factory's labour force comprises 1,200 employees, of whom 750 are direct labour. The assembly department employs 250 people, including 40 on each production line.

Seventy percent of the labour force is semi-skilled and the majority of these are males.

When the experiment started (April 1973), it concerned 42 people on two production lines and by the end of 1974 it involved 75 people on three lines.

Background

When the experiment started, production was on conventional assembly lines with a length of at least 50 metres. The job cycles were short, viz. about 1½ minutes.

The personnel were tense and there were latent feelings of dissatisfaction. Absenteeism figures were high for male employees and production was below target.

The social climate was poor. There was a great deal of tension and an atmosphere of social conflict.

Work structuring created a possibility of bringing about a change in this situation.

Other initiatives had been taken in the field of job analysis, remuneration systems, safety and training but had proved inadequate.

Content of Change

Inspired by the work structuring idea, the manager of the TEO department proposed the introduction of a new method of working with new products.

Small work groups and jobs with longer cycles (5–10 minutes) were made up in order to reduce the number of employees on the production

*Editor's Note: This and the following examples of projects have been excerpted from Chapter 3.

lines and to remove the drawbacks which are inherent to this type of organization, such as monotony.

Three people were given the job of building a "washing block". The entire job consisted in assembling and inspecting a whole unit. This made it possible:

for every member of the group:

- to work at his/her own pace;
- to do a larger job, with more operations;
- and, in some cases, to learn all the jobs in the group.

for the group:

- to undertake new tasks such as inspection, repair, drawing supplies and accounting;
- to make a choice and decide about the distribution of jobs in accordance with each other's preferences;
- to teach each other the successive operations.

As a result of this set-up everyone is more familiar with the results and feels more involved. The fact of sitting round the same table makes mutual contact easier. Generally speaking, the effect is to increase flexibility and a sense of responsibility.

Evaluation

This new work organization was not made in response to any specific demand from the personnel, but was rather the initiative of a staff department which launched the innovation in the course of doing its own job. Had the plan failed, the result would have been a return to the original situation. The most positive aspect of the affair was that people seized the opportunity to try out something new.

Although the new set-up could not be more expensive than the old one, new equipment had to be developed and the resultant technical problems had to be solved.

More floorspace was needed and an immediate increase in productivity was not expected. One of the aims of the experiment was undoubtedly to involve the personnel more actively in the organization.

Consultation in the initial phase was on a large scale, co-operation with the operators being indispensable if any real change was to be made. Relations between the personnel and those in charge of the experiment were good. Lower management, on the other hand, adopted a reserved attitude and waited to see what would happen. Their task in the operation was indeed discussed. Their reticence became a problem which had an adverse effect on the results of the project.

Conclusions

This type of experiment creates favourable conditions for the development of participation but is not sufficient to bring it about in fact. We wanted to find a general answer to the personnel's lack of interest without first conducting a study into the needs and the causes of their dissatisfaction. Under such circumstances how can one judge the results of the project in this respect?

Democratic procedures were created. Were there any special expectations underlying the decision to do this?

The result is that those concerned can now determine the limits of the organization; their capacity for participation has, one might say, been tested.

The number of possible choices has, in fact, been increased. Employees have been consulted about certain aspects of the situation, but without involving them in the project as a whole.

This case study shows that participation is only a technique, a secondary aid, as long as it is limited to one category of personnel or to only a part of the mutual relationship.

Finally, structural changes have been carried out and the problems of lower management studied, a process which started at the end of 1974 and is being continued. In view of the success of the experiment possibilities now exist for extending it.

BLACKBURN, ENGLAND

Autonomous Group Working in a Wire Division at Mullard

Situation

About 4,200 people are employed at the Blackburn factory complex, of whom about 1,100 are "white collar" employees. The Wire Division has 483 employees, of whom approximately 90 are "white collar". Tungsten and molybdenum wire, electrodes and spirals for electric lamps are produced there. The Division's products are then used to produce a variety of wire components and sub-components, mainly for the Hamilton lamp factory, other concern factories and third parties.

The technology involves a fairly high degree of expertise and the utilization of high-speed wire coiling equipment. A number of female operators are employed in the final inspection department for lamp spirals, part of which is concerned with the "gettering" and final inspection of spirals for vacuum lamps. These female operators were fairly isolated geographically and had monotonous, quite short cycle (6–10 minutes) jobs to do. It is in this area that group working has been introduced.

Background

The project was started in January 1973. Previously, production methods involved 10 sequential operations carried out by individual operators with little or no task interaction. The project was instigated by the Divisional Manager, who wanted to test out the benefits of "work structuring" in his own area of the organization. In another part of the factory complex, autonomous group working had been developed, by participative working, in an attempt to create more meaningful jobs and more group cohesion—and by doing so to reduce absenteism and labour turnover. Thus, some experience of "team working" was already in existence.

Content of Change

The local Personnel and TEO officers were the consultants to the group of seven hourly-paid female operators who were invited to place themselves in an evolving group situation in which it was hoped they would contribute to increasing the degree of autonomy available to them. One objective of the group is to maintain a balance between individual abilities and the scope of change which is made possible by the technology and the organizational climate. The seven operators are now working as an autonomous group, and are collectively responsible for producing the required production programme. Previously, they had taken part in an educational process which covered many social and technical aspects of "work structuring" and had then made their own decisions concerning how they would structure their work, as a group.

Evaluation

Prima facie savings are satisfactory. Data on such variables as output, quality, labour turnover and absence are difficult to interpret due to a) the small size of the group and b) the lack of a control group for comparison. However, although the level of output is subject to fluctuation, it is generally higher within the group than in the "before group" situation. A report which attempts to identify the social aspects of the situation is currently being prepared, and is based upon a series of informal interviews, by the consultants, with group members. The feedback from participants tends to show that the advantages of team working outweigh the disadvantages—"Group working is better", "Not being bothered by a charge-hand helps" and "I enjoy the sense of responsibility and achievement from seeing a product through from beginning to end"—although the team still has interpersonal problems and a perceived lack of support at times—"I like working in a group better, but it will only work when people have an equal chance of doing different jobs", "There would be more variety if the group was properly organized" and "Must make sure supervision shows a positive interest in the idea and the jobs from the beginning".

Interview topics and data which have been fed back to supervision by the consultants for discussion include

a. the commitment of supervision;
b. the need for a common approach by different levels of supervision;
c. the need for more training—e.g. to facilitate job rotation, in the pre-group situation;
d. the need for consultants to know when to intervene in/withdraw from the management situation;
e. the level of autonomy to be offered to the group—e.g. with respect to planning, communications and changes such as computerisation of some processes.

Conclusions

In retrospect, there is a need for management and consultants to be more prepared to consider, in depth, the implications of such a change process as for example, their own roles and relationships, the power structure implications of group working, and the long-term policy of the organization vis-à-vis this philosophy and approach. There is also a need to recognize that a new group must be given a *chance of survival*, for example, by taking away some of the normal production pressures at the beginning of the change process, and that spreading the learning from such a process is a complex matter requiring much energy and commitment. The group is continuing to work satisfactorily, in socioeconomic terms, but future plans for the expansion of group working in the Division are uncertain and are subject to many differing pressures from both management and trade unions. This is—from the consultant's viewpoint—because of the lack of coordinated policy on work restructuring.

MONZA, ITALY

An Approach to Work Structuring in a Video Factory

Situation

The Monza Video Factory produces color TV sets, Tinyvision sets, tuners and components.

A total of about 1,200 direct workers are employed in this factory. Production is based on traditional assembly lines, job cycles are rather short and jobs well defined and related to function groups with relevant payment scales. The piece-rate system is still applied.

Background

What created the need for work structuring?

1. People now working in the factory have higher expectations regarding their jobs. Since personnel turnover is very low (about 2–3%) and employees' ages and years of service are increasing, something has to be done to provide more interesting jobs.
2. It is to be expected that people taken on in the future will be better educated and will accordingly want jobs adapted to this higher level.
3. The new national labour contract sets out the need for workers to be promoted to higher categories and for the job content to be adapted by providing a wider range of facilities and responsibility.
4. Within the context of the national labour contract the detailed contract at company level provides for activities in the field of work structuring.

Therefore steps were taken to study the possibilities of changing working conditions.

Approach

A steering group was installed in which the following were represented: Video Management, Personnel Dept. of the Monza plant, Video TEO, Central N.O. Personnel Dept. and N.O. TEO.

Steering group objectives:

To carry out a preparatory study of previous workstructuring experiments within the Concern in order to determine the best approach.

To work out a proposal for the Italian CCD concerning the steps to be taken and the relevant financial consequences.

After an extensive study the steering group arrived at the following guidelines which formed the basis for further steps:

1. The aim should be to promote as far as possible the creation of autonomous groups and stimulate job enrichment in the event that autonomous groups are not feasible.
2. The limits of responsibility of each group should be clearly defined and formulated.

3. Assembly should be divided into functional modules which could be considered the finished product of individual autonomous groups.

4. Each functional module should comprise the following phases: assembly, inspection, repair and responsibility for final quality. (The supply of materials and components are not included within the scope of responsibility).

5. The group would work best if the number of people in it was between 7 and 10.

6. Measures should be taken to train people so that the workers within the group are interchangeable. Group processes should also be stimulated.

7. The quantities produced by each group must be proportional to the number of working hours worked by each group (i.e. in the event of a group member being absent the quantity produced would decrease if a replacement were not provided).

8. The payment system would be changed from the existing piece-rate system to group payment taking into consideration the following:

—individual basis : age, number of years with company, job level;

—group premiums : quantity and quality.

To work out detailed proposals based upon the above guidelines a working party was formed in which the following functions were represented: Video Factory Personnel Department, Video TEO, Methods, line management from relevant departments, and a specialist project leader from Central N.O.

The following overall schedule was worked out:

1. Project preparation by working party
   —departments in which the experiment with pilot autonomous groups would be initiated
   —products and/or functional modules affected
   —establishment of pilot groups
   —estimates of cost and time needed.
2. Final proposal formulated by steering group and working party.
3. Presentation to Factory Management.
4. Presentation to CCD.
5. Presentation to local union representatives and open discussions with people involved.
6. Formation of pilot groups and start of training.

7. Practical steps:
   —Adaptation of lay-out.
   —Installation of necessary equipment.
   —Preparation of working methods and relevant instructions.
   —Time studies.
8. After agreement with Factory Management, start of the experiment.

The project has now arrived at the stage of presentation to local union representatives.

Detailed documentation explaining the project has been compiled for this purpose.

It is intended to start provisionally with a functional module in each of four departments, and two pilot autonomous groups in each module.

Conclusions

1. The whole project is still at too early a stage for results to be presented.

2. The preparatory study preceding the experiment revealed that the financial consequences are fairly considerable, mainly due to the extra space needed, the adaptation of the assembly lines and changes in layout, extra equipment, and extra time needed from specialists in various staff departments.

3. It is not possible to calculate any definite profitability justifying the necessary investment and initial cost.

4. The high financial consequences have to be justified by greater job satisfaction, eventually resulting in decreasing absenteeism and an easing of social tensions, as well as a more balanced relation between wage level and job performance.

EINDHOVEN, THE NETHERLANDS

Organization Development in a Mechanical Workshop of the Research Laboratories

Situation

The project is being conducted in one of the four mechanical workshops of the Research Laboratories in Eindhoven, the task being to support research by developing aids, instruments and techniques. About 80 skilled workers (mainly milling-cutter operators and turners) and 10 supervisors are employed in the workshop.

Background

Increasing criticism was being directed at the Methods and Costing Department, where employees saw definite restrictions on their responsibilities. Furthermore the existing philosophy of piling up units and adjusting the rate and piece-work systems accordingly, seemed to have an adverse influence on their attitude to work.

Finally, a growing number of problems resulted, presumably from the fact that the majority of the skilled workers—having left school with great expectations for the future—had started to realise four or five years later that the chances for further promotion were extremely limited.

Content of Change

Inspired by the results of other studies, management developed a plan to introduce group work. However, before putting the plans into action it was decided to find out more specifically the ideas of those involved about the problems. Elections were held to set up a *work group* representing all levels in the department. The final opinions of the group were to be submitted to a *steering group* which was set up at about the same time and which comprised management representation, a representative of the Personnel Department and a member of the Works Council. Both groups included an organization-development consultant. The opinions of those involved showed—unexpectedly—that it was not the introduction of group work, but the creation of a trustful relationship between management and workers which was given first priority.

After various discussions and exchanges of ideas between management and workers it was decided to establish *a permanent consultative body* for group work consultation . A number of matters such as career planning and job rotation were also taken up. Finally, the decision was taken to make an analysis of the technico-organizational aspects of the department. This study, for which capacity equivalent to one man-year has been set aside, is intended as a basis on which to build up group work, i.e.:

creation of small groups to be responsible for a specific task, including preparation, task distribution and checking;

creation of direct contacts between the skilled worker and the researcher to decide by joint consultation what task is to be carried out and how it is to be done;

restriction of the traditional supervisory functions and gradual delegation of responsibility to those directly concerned with performance of the work.

Evaluation

As far as the economic results are concerned, an improvement in the care of machines and tools and better quality have been noted, probably as a result of better motivation. A shift in the "work package" has already taken place. More and more skilled workers have gone over to working directly for the researchers and have shown much greater concern for the contents of the orders and the way in which they are executed.

The opinions of the skilled workers, supervisors and main customers regarding the progress made were obtained by means of interviews. These showed that the original distrust has practically disappeared. Almost everyone agrees that relations with management have taken a turn for the better.

The foremen have become much more open and the basis for confidence has grown. Half the skilled workers regard the work consultation group as a useful channel. Supervisors, customers and skilled workers are all in favor of close contact with the researchers.

Conclusion

We are dealing here with an intensively supported project which aims at maximum participation by those concerned. Priority has gradually been given to *the restoration of relationships* that had worsened.

The advantages of this are that the project has become a matter for the organization itself and that the organization has gone through a process of *self-renewal*.

A second notable characteristic of this project is that the approach is consistently to the department *as a whole*. Small experiments have been provisionally turned down. It is realised within the department that it is easy enough to set up one or two autonomous groups. *Experience in other large workshops within Philips has shown that the difficulties begin when a decision as applied to a whole workshop:* problems centering on the training of workers and lower management, relations between groups, the adjustment of machines, procedures etc. begin to arise.

On the basis of the confidence which has been created, it is hoped satisfactory solutions will be found by means of consultation and a thorough analysis of the organizational and technical aspects.

COPENHAGEN, DENMARK

Organization Development at Philips Radio A/S

Situation

The factory is comparatively small—about 450 direct workers (mainly female) of whom 150 are indirect. The production is highly specialized on two different items—channel selectors for television receivers (mass production) and professional television equipment for laboratories, TV factories and TV studios/transmitters, and equipment for medical purposes (series production).

Background

In 1969 it was decided to start an organization development programme, comprising in the first instance *experiments with semi-autonomous groups*, but also with a view to *changing the organization* from the present mechanical system (bureaucratic organization) to a more organic system (group orientated).

The purpose was:

1. to make work more attractive and so increase interest in the job;
2. to utilize the resources of the people employed and so achieve higher efficiency.

In the first place we expected that the programme would reduce labour turnover, which at that time was rather high. Secondly, we thought that to attract young people who even at school nowadays learn to work in groups and to decide for themselves, we would have to change our organization in the future to meet their requirements.

We also felt that our existing staff could gain from such a change and that the entire organization would work more efficiently if the resources of the employees were better utilized. Our belief was based mainly upon studies by McGregor (X-Y theories) and Herzberg (motivation theories). As better understanding of human behaviour was involved, we also realized that an external consultant could be of great help and a psychologist from the Danish Technological Institute was invited to join the project.

Starting Up the OD Programme

The *"unfreezing period"* started with a conference in which 64 people from all levels of the organization participated and during which the basic theories regarding organization development were outlined. Then a smaller group of 16 people was formed which met weekly for six months and discussed the possibilities of using organization development. Finally it was decided to form a small Steering Committee consisting of the factory manager, the chief shop steward, a shop foreman and the psychologist, with the task of informing employees about organization development, collecting information about the experiments and coordinating and supporting the experiments.

At a later stage the psychologist left the Steering Committee, which was eventually expanded to include two other staff members and an indirect employee who—after adequate training—could act as a "change agent" in addition to his normal job in the organization.

Conferences for direct and indirect personnel were organized to disseminate knowledge of organization development and several people followed external courses on the subject.

At the same time interest in new forms of management was growing in the organizations within the Iron and Steel Industry. The Employers' Association and the Trade Unions therefore invited enterprises to participate in experiments, mainly involving semi-autonomous groups in order to acquire experience in this field.

Seven factories were chosen—we were one of them—and in this way we also received official support from the above organizations to conduct the experiments.

Content of Change

As mentioned above, we have started a complete OD-programme resulting in:

1. *Semi-autonomous groups* in the production departments;
2. a special effort to introduce *OD in the laboratory organization*, and
3. substantial changes in the daily co-operation within the *organization structure*.

The details are as follows:

1. *Semi-Autonomous Groups.*

Semi-autonomous groups have been formed both in the mass production and the series production departments and in ancillary departments:

a. channel selector department: four groups, seven to eight female workers each;
b. assembly of measuring equipment: 35 female workers and two male workers + three male workers in adjacent store room;

c. hardware production department: one group consisting of three to six male workers spotwelding frames for CTV set K8 (this group was dissolved in Spring 1973 when production ceased);
d. maintenance department (buildings and installations): whole department consisting of three skilled workers and seven unskilled workers.

The semi-autonomous groups in the production departments have abandoned the traditional belt system and, instead they now occupy individual positions with a buffer stock between each position, so that group members can work at their own individual speed during the day. The group is responsible for organizing its daily work and job rotation is used to a very great extent. Where convenient, the group members contact ancillary departments themselves so that the job of the supervisor has changed from normal leadership and supervision of the group to a more consultantlike job.

It has proved necessary to abandon the piece-rate system of payment and to adopt a fixed-wage system which guarantees the income for a certain period (three months) but which is still related to the output of the group during the previous period.

The semi-autonomous group system has resulted in increased interest in the product and in the work in general. Quality is consequently higher while less absenteeism and turnover of personnel have been noted. The output of the group is the same as with the normal belt system but in view of the improvements mentioned the overall efficiency must be considered superior to that of the belt system, even if the investment in tools and jigs is somewhat higher.

Above all, members of the semi-autonomous groups maintain that they do not want to return to the normal assembly belt system, because they feel more satisfied under the new working conditions.

In the maintenance group very interesting results have been obtained. People employed there were formerly set to work by their supervisor (a machinist) with the result that, generally speaking, they were not very interested in the job, worked slowly and "hid" so that the supervisor could not find them when needed.

In the new situation with the semi-autonomous group, incoming orders are placed in a box at the supervisor's desk and the group members, on finishing one job, go and look for the next job themselves, while the supervisor acts mainly as a consultant and does not intervene except in emergency cases.

The whole group feels a joint responsibility for the work of the department, they help each other, and hence the efficiency of the department has increased considerably.

2. *OD in the Laboratory Organization.*
In January 1972 the entire PIT Laboratory (45 people) followed a one-week organization development course in order to create a basis for further job delegation. This was followed by one-week's sensitivity training for each member of the leader group.

The result is that the former, traditional, autocratic style of leadership has gradually changed to a system which rather resembles "Management by objectives".

The laboratory staff are much more involved in the job and have a greater feeling of responsibility. The interest in activity as a whole has increased and, among other things, close contacts have been created between the development engineers and the direct workers in the assembly department. A remarkable decrease in the turnover of young engineers has also been observed.

The alignment group of the measuring equipment section was formerly organized as a separate group but has now been made an integral part of the laboratory, providing much better contact between the highly skilled technicians in this group and the laboratory engineers. This has resulted in an extremely marked increase in efficiency.

3. *Organization Structure.*
The structure of the organization has not changed much but better mutual understanding and co-operating among individuals and departments have developed within the normal framework. These have been brought about partly by courses and conferences and partly by using approaches such as: team development, confrontation meetings, role analyses and data feedback. Decisions are no longer made solely at top level but attempts are made to reach them at the level where the information is available. More important decisions are made by consultation with everyone involved. This takes more time but when the decision is made it is accepted by everyone and carried out without obstruction.

During the past few months, however, a number of "horizontal groups" have been created in the organization; for instance, the heads of the three channel-selector manufacturing departments, the production engineering department and the factory laboratory have formed a leader group in which they assume joint responsibility for the total production. Thus a former tendency to consider only the efficiency of one's own particular department has been reversed to the benefit of the factory as a whole.

Product groups with members from the various departments have been formed for the various types of channel selectors in order to solve in the most appropriate way all the problems that arise, i.e. a clear shift is taking place towards a group-oriented organization.

Conclusion

It has been proved that the introduction of an OD programme can improve co-operation within, and the functioning of, the organization. It is a slow process, however, because it depends upon a change in the behaviour of people in the organization. When you have been accustomed to the normal autocratic manner of leadership for years, you simply cannot adapt overnight to the demands of group-oriented organization—it takes years.

Remarkable results can be obtained even in the short term with semi-autonomous groups, but it should be noted that this will inevitably change the job of the supervisor, who will take over more boundary work and thus influence the job of his superior, and so on. Therefore large-scale introduction of semi-autonomous groups in an enterprise demands radical changes in the organizational set-up.

The introduction of an OD programme depends upon the existing "culture" in the particular organization. The programme must therefore be "tailor made" and there are no fixed rules for doing this. Consequently evaluation is based upon a trial-and-error method, and it is important that employees in the organization should be able to accept the inevitable failures and adapt the programme accordingly when failures occur.

PART V:

SOCIOTECHNICAL SYSTEM STUDIES:

WHAT DO WE REALLY KNOW?

Introduction

Although a great deal has been written about sociotechnical system theory, diagnosis, and change, much is still unknown. To discover the current state of the art, we must study the research that has been done, make comparisons, and delineate those areas that are not clear. The articles in this section have all provided partial answers to the question "Sociotechnical systems: what do we really know?"

Cummings reviews sixteen reports of sociotechnical interventions, identifies common characteristics, and compares results. He finds that most of the interventions involved changes in organizational reward systems, employee autonomy, technical or physical work methods, task variety, information and feedback, and interpersonal and group processes. He notes that most of the experiments took place in male, blue-collar, unionized organizations and that most reported positive results.

Miller revisits the site of some research A. K. Rice did sixteen years earlier in order to assess the permanency of the changes that were introduced at that time. Miller finds that market changes, difficulties in processing materials, and turnover have led to regression of the changes in some cases, but that in other cases the changes have withstood similar pressures. His discussion of the changes that occurred is of great interest for those who wish to establish durable organizational change.

Pasmore reports the results of a field experiment designed to compare the effects of sociotechnical system, job-redesign, and survey-feedback interventions in comparable units of an organization. He reports that although the interventions did not differ in terms of their impact on employee attitudes, only the sociotechnical system intervention resulted in major productivity improvements and cost savings.

McCuddy points out areas for future research by sociotechnical system theorists, including the need to define and measure technology more adequately; the need to delineate social systems and their characteristics from other components of an organization; the need to understand the relationships between the social and technical systems of organizations in terms of their joint and separate influences on organizational behavior; the need to assess more fully the impact of different sociotechnical arrangements on organizational performance; and the need to find better ways to conduct research on sociotechnical systems.

The authors in this section indicate that there is still much to be learned about sociotechnical theory and intervention and they provide stimulating challenges for the future. But, at the same time, these studies are evidence of the progress we have made in understanding how to improve organizations. In the next section, several noted experts provide their own perspectives on sociotechnical system theory and its development and application.

Sociotechnical Experimentation: A Review of Sixteen Studies

Thomas G. Cummings

Attempts to improve both productivity and the quality of work life are gaining acceptance in contemporary organizations. The push toward work improvement can be seen in a number of industrialized countries in which advanced technologies and sophisticated work forces place heavy demands on traditional work designs. Norway and Sweden, for example, have begun to expand democratic principles to the shop floor. They have made major commitments to industrial democracy. On a smaller scale, several organizations in Canada, Great Britain, and the United States are experimenting with innovative work structures. Using a variety of approaches, these firms have shown that work can be designed for improved economic and social outcomes (Cummings & Molloy, 1977).

As studies of work improvement have been described in the scientific and popular literature, the term "sociotechnical systems" has come to characterize many of the experiments. From its British origins at the Tavistock Institute of Human Relations (Trist & Bamforth, 1951), the sociotechnical approach to work design has done much to integrate and systematically organize people and technology for productive achievement.

The sociotechnical perspective furnishes a conceptual and practical base to enhance both the social and economic outcomes of work. Although research results have been generally favorable, there have been few attempts to review the empirical literature systematically. Typically, sociotechnical studies have been summarized in a cursory manner that limits the reader's understanding primarily to organizational changes and their overall results. When specific experiments have been reported in depth, there has been wide variation in the concepts and data presented, making it difficult to compare different studies or to isolate the studies (Cummings, Molloy, & Glen, 1975; in press).

The purpose of this paper is to examine sixteen selected sociotechnical studies on a number of common dimensions having to do with organizational changes and their results and characteristics of workers and organizations in the hope that a systematic report will provide a clearer understanding of the sociotechnical approach and furnish empirically based knowledge for experimentation. The sixteen experiments reported here represent only a small portion of the sociotechnical projects that have been carried out to date. Many additional studies were excluded either because they were unavailable to the author or because they provided insufficient data for a review of this kind.

ORGANIZATIONAL CHANGES AND RESULTS

The sixteen sociotechnical studies chosen for this review involved a number of organizational changes to improve productivity and worker satisfaction. A majority of the studies were concerned with the following types of improvements:

1. *Pay/Reward Systems*. The most frequently mentioned change in pay/reward systems was the introduction of a group pay scheme, such as a marginal group bonus or a common pay note for all members. This usually coincided with a movement from individual to group forms of work. For example, workers who had previously been paid differentially according to their individual job classifications were given a common group pay rate.

2. *Autonomy*. The delegation of authority or discretion downward in the organization was an important element in most of the experiments. The primary focus was to provide group members with task autonomy. This enabled workers to deploy themselves in a variety of ways to match changes in their work environments. Generally, the studies reported increases in discretion in reference to autonomous group formation. This focus on group effort follows directly from the systems orientation underlying the studies. Experimenters were concerned with the systemic relations among interdependent tasks and the formation of discrete work systems to coincide with these natural task boundaries.

3. *Technical/Physical*. Technical changes and modifications of the physical work setting were a major feature of the studies. This is not surprising in light of the attention given to both the social and techno-

logical parts of work by sociotechnical experimenters. Technological changes frequently involved the physical relocation of equipment and materials to form relatively independent task groups. A variety of smaller, more subtle modifications having to do with tool improvements, material handling, and work flows often accompanied these more obvious technical changes.

4. *Task Variety*. Increasing the number of tasks workers could perform was a prerequisite for designing autonomous work groups. This provided workers with the skills and knowledge to control variances from goal attainment within the work system. It also gave team members considerable flexibility for rotating among different tasks and for balancing their work loads over a variety of emergent conditions.

5. *Information/Feedback*. Many of the experiments modified the flow of task-relevant information to workers. This usually concerned the provision of more timely and relevant knowledge of the group's performance, such as quantity and quality of production. In those cases in which individual jobs were regrouped into work teams, the feedback system was enlarged to include information about group performance.

6. *Interpersonal/Group Process*. The formation of work groups invariably entailed changes in the patterns of contact among individuals. Members had to learn new skills as well as develop new work relationships. Although much relationship building was a naturally occurring process, several of the studies reported explicit team-building activities, such as group meetings with an outside facilitator, supervisor/subordinate goal setting, and intergroup problem solving.

The changes mentioned above are presented in Table 1 for each of the sixteen research studies. Data about the productivity and attitudinal effects of these changes are also included. The most prevalent change studied was the amount of autonomy or discretion given to workers (fifteen of sixteen studies). This was followed by modifications in interpersonal/group process (twelve studies), information/feedback (ten studies), task variety (ten studies), technical/physical (ten studies), and pay/reward systems (nine studies). The results of these changes were overwhelmingly favorable. Of the fifteen studies measuring productivity, all but one showed an improvement. Similarly, of the nine studies measuring attitudes, only one reported a totally negative result. Although the studies' weak research designs lower the validity of these findings, sociotechnical experimentation seems to be a promising strategy for improving work conditions.

The majority of the experimental changes involved autonomous group formation. Research has shown that at least three conditions are necessary for such groups to function effectively: (1) a relatively whole

Table 1. Organizational Changes and Results

CODE:
Blank = not changed or measured
x = variable changed
+ = variable increased
− = variable decreased

	Pay/Reward Systems	Autonomy	Technical/Physical	Task Variety	Information/Feedback	Interpersonal/ Group Process	Productivity	Attitudes
Bregard et al. (1968)	X	X		X	X	X	+	+
Cummings & Srivastva (1977a)	X	X	X	X		X	−	±
Cummings & Srivastva (1977b)		X	X	X		X	+	−
Emery et al. (1970)	X	X		X	X	X	+	+
Englestad (1970)	X	X	X	X	X	X	+	
Gorman & Molloy (1972)		X		X	X	X	+	+
Prestat (1971)		X	X		X	X	+	
Barry Corporation (undated)	X	X	X		X		+	
Rice (1958a)	X	X				X	+	+
Rice (1958b)		X	X		X		+	
Trist et al. (1963a)	X	X			X	X	+	
Trist et al. (1963b)				X		X	+	
Van Gils (1969)		X	X		X	X	+	+
Van Liet (1970)		X	X	X			+	+
Vossen (1969)		X	X	X				±
Walton (1972)	X	X	X	X	X	X	+	

task; (2) control over environmental relationships; and (3) control over task activities (Cummings & Griggs, 1977). The changes studied provided these conditions. Increases in autonomy gave team members control over their task and environmental exchanges. Individuals were able to decide how to perform the task and how to relate to relevant segments of the environment, such as maintenance, purchasing, and sales. Interpersonal/group-process interventions provided workers with the social skills necessary to solve a common task. The focus was on interactions among team members as well as on relationships with other groups. Changes in information/feedback furnished workers with the information necessary for goal-directed activity. Relevant and timely feedback enabled individuals to adjust their performances to match changes in the task and environment. Increased task variety provided workers with the requisite skills to perform their tasks. This reduced members' task dependence on others external to the group. Technical/physical modifications formed relatively whole task boundaries. Equipment, materials, and work flows were rearranged so that interdependent tasks could be performed within a definable physical boundary. Finally, changes in pay/reward systems reinforced group task structures because members were rewarded as a group.

The combination of these changes provided workers with the necessary conditions for autonomous group formation. Most studies reported that the experimental changes were implemented as intended. The large number of positive results attests to the success of the change programs. However, three of the four studies in which the changes were only partially implemented also showed some form of negative result (Cummings & Srivastva, 1977a, 1977b; Vossen, 1969). In the Cummings and Srivastva experiments, for example, the company was experiencing a rapid shift in the market for its products. Attempts to cope with this environmental change resulted in a good deal of internal disruption that intruded on the experimental process. Rather than stop the experiment, the company decided to partially implement the conditions for autonomous group effectiveness. This meant that workers had only limited control over their tasks and environments. Failure to implement fully the necessary conditions for autonomous group functioning may lead to a work structure that thwarts members' ability to be responsibly autonomous.

CHARACTERISTICS OF WORKERS AND ORGANIZATIONS

The sixteen studies were carried out in a variety of organizations with different types of workers. Much of this contextual information was not reported systematically across experiments. This restricted the comparative data to only seven standard items: (1) sex of workers; (2) occupational status (blue or white collar); (3) union membership; (4) type of work restructured; (5) number of experimental subjects; (6) whether workers participated in the change program; and (7) country in which the organization was located. Although these categories were derived from the data available in the studies rather than from a preconceived theoretical classification, they reveal much about the workers and organizations involved in sociotechnical experimentation.

Table 2 presents the contextual information for each of the experiments. From those studies that reported sufficient data about the employees involved, it appears that sociotechnical experimentation is oriented primarily toward male, blue-collar, unionized workers. The types of work restructured seem consistent with this finding. In all but two studies (Cummings & Srivastva, 1977b; Gorman & Molloy, 1972), workers were required to relate directly to technology to transform raw materials into finished products. Chemical processing, machining, packaging, wire drawing, weaving, coal mining, key punching, and assembling are predominately production oriented. These tasks, with the possible exception of packaging, key punching, and assembling, are typically carried out by male, blue-collar employees. The number of individuals involved in the experiments ranged from seven to 350, with a median of 41.5. All studies that reported worker involvement in the change program revealed that employees participated directly in planning and implementing the organizational changes. The countries in which the experiments were carried out were mostly Western industrialized nations—the United States (four); the Netherlands and Norway (three each); Great Britain and India (two each); and France and Ireland (one each).

The contextual data suggest that sociotechnical experimentation has been prevalent in the industrial sector. Indeed, most of the studies were carried out in shop-floor production systems in which people and technology interacted to perform work. The production orientation of these studies may be explained in reference to sociotechnical system theory, one of the few perspectives that integrates social science and engineering. The social and psychological dimensions of work are explained in terms of engineering and the way the technological system as a whole is related to the work environment (Trist & Bamforth, 1951). Given this link between social and technical systems, it is not surprising that industrial research tends to focus on production situations in which the worker and technology interface is most significant.

Service-oriented forms of work, on the other hand, depend primarily on people rather than on technology for performance. In this case, work takes on the characteristics of a psychosocial system in which the

Table 2. Characteristics of Workers and Organizations

	Sex	Occupational Status	Union Membership	Type of Work	Number of Workers	Participation in Change Program	Country
Bregard et al. (1968)	?	Blue	Yes	Chemical	40	Yes	Norway
Cummings & Srivastva (1977a)	Male	Blue	Yes	Machining & packaging	43	Yes	U.S.A.
Cummings & Srivastva (1977b)	Male/ Female	White	No	Die designing & estimating	28	Yes	U.S.A.
Emery et al. (1970)	Male	Blue	Yes	Wire drawing	12	Yes	Norway
Englestad (1970)	Male	Blue	Yes	Chemical processing	28	Yes	Norway
Gorman & Molloy (1972)	Female	White	?	Clerical	19	Yes	Ireland
Prestat (1971)	Male	Blue	?	Weaving	100	Yes	France
Barry Corporation (undated)	?	Blue	?	Assembling & packaging	350	?	U.S.A.
Rice (1958a)	Male	Blue	Yes	Automatic weaving	28	Yes	India
Rice (1958b)	Male	Blue	Yes	Nonautomatic weaving	66	Yes	India
Trist et al. (1963a)	Male	Blue	Yes	Coal mining	41	Yes	Great Britain
Trist et al. (1963b)	Male	Blue	Yes	Coal mining	41	Yes	Great Britain
Van Gils (1969)	Female	White	?	Key punching	66	Yes	Netherlands
Van Liet (1970)	Male/ Female	Blue	?	Assembling	7	Yes	Netherlands
Vossen (1969)	?	Blue	?	Assembling	20	Yes	Netherlands
Walton (1972)	?	Blue	?	Processing & packaging	350	?	U.S.A.

psychological aspects are best understood in reference to the social dynamics operating in the work place. Sociotechnical theory is not as directly applicable for service-oriented work.

The production focus of the experiments is also consistent with the early development of the sociotechnical approach. Starting with the coal mine (Trist et al., 1963a, 1963b) and weaving (Rice, 1958a, 1958b) studies, the theoretical and applied bases of this approach have had a production focus. Most of the initial studies were in production systems, and the development of the sociotechnical approach has followed a similar orientation. For example, two analytical models have been devised for designing work systems (Foster, 1967). The method applicable to production systems is relatively more developed and tested than the service-system model. Although the sociotechnical perspective originated primarily from experimentation with production systems, there is no reason to believe that it cannot readily be extended to other forms of work. Experimentation in other settings is a promising path for future sociotechnical development.

A second distinguishing feature of the sixteen studies is worker involvement in the change process. This originates from both practical and value-oriented considerations. Pragmatically, worker participation in the experiment reduces the likelihood of resistance to change (Coch & French, 1948). When employees have some say about the conduct of the experiment, the changes considered, and the methods of change, they are more committed to implementing the innovations. This is because the very process of participating in change allows workers to meet needs they value, especially those related to autonomy, recognition, and challenge. Hence, participation is a path to need achievement.

Employee cooperation also is wise in terms of diagnosis and redesign of the work system. Those individuals who are directly involved in the day-to-day operation of the work unit possess intimate knowledge of its dynamics. Their participation can provide more relevant and valid data for diagnosis as well as proposals more tailored to the situation.

The value implications of worker participation in organizational change stem from a belief in democratic ideals. It is no coincidence that sociotechnical experimentation has been associated with the movement toward industrial democracy (Emery & Thorsrud, 1969). Briefly, the general objective of industrial democracy is to extend the principle of government by the people to the work place. Although this goal typically is seen as congruent with the democratic ideal, there has been wide variation in how it is achieved. European experience traditionally has involved worker representation on managerial boards. This has done much to redress any imbalance of power in the firm, but it has failed to achieve democratic handling of day-to-day problems on the shop floor. To overcome this difficulty, a few countries (most notably Norway and Sweden) have redefined industrial democracy to include worker responsibility for decisions affecting the job. Attempts to achieve this rank-and-file objective have frequently followed a sociotechnical strategy. Workers are included in every phase of the change program, and if employees are unionized, union sanction is seen as a prerequisite to experimentation.

A final characteristic of the studies concerns their cultural context. As noted, most were carried out in Western industrialized countries. The cultural and economic conditions in these nations seem to support work-improvement experiments along the lines reported here. As discussed, a belief in democratic ideals is congruent with a push toward worker participation in decisions affecting work life. Although the countries involved vary in the extent to which they support worker democracy publicly, their democratic traditions lend support to such experimentation.

The countries also appear to be undergoing certain social and economic changes that make sociotechnical experimentation desirable. The social conditions that give traditional forms of work their purpose and dignity are gradually being transformed. Life styles and values affirming freedom and self-expression exist in conjunction with customary values of self-discipline and hard work. An accumulation of social and economic changes has forced many industrialized countries to experiment with alternative forms of work. The sociotechnical studies reviewed here represent this search for work structures with positive social meaning and economic validity.

CONCLUSIONS

The sixteen studies reviewed follow a distinct theoretical orientation in which both the social and technological aspects of work are placed in a systems framework, drawing attention to the interface between people and technology and furnishing a much-needed integration of the social sciences and engineering. By placing work in the context of the wider environment, sociotechnical research extends the focus to those external forces affecting task performance.

The theoretical base of the research has a subtle influence on the unit of change. Rather than concentrating on individual jobs, sociotechnical researchers examine the more fundamental issue of how tasks are interrelated systemically. This often leads beyond single job boundaries to encompass task groupings, structured so that members can control both task activities and environmental exchanges. This provides the work system with the necessary autonomy for self-regulation and furnishes workers with opportunities for freedom and growth.

The change strategy used to implement sociotechnical designs focuses on limited parts of the total organization. Pilot projects that involve small numbers of employees are carried out with fuller worker participation. This reduces the likelihood of resistance to change and allows a more realistic diagnosis of the work system. Worker participation in the change process is also a direct outgrowth of the democratic values shared by many sociotechnical researchers.

Insofar as it is possible to describe a typical sociotechnical study, it includes male, blue-collar, unionized workers who perform industrial tasks. These employees number about forty, and they participate directly in the change program. The experimental changes involve forming autonomous work groups; specific modifications include increases in autonomy, interpersonal/group processes, feedback of results, and task variety, with parallel changes in the technical/physical setting to form whole task groups and in the pay/reward system to reinforce group performance. If the necessary conditions for autonomous group functioning are implemented, the likely results are increases in productivity and worker satisfaction.

This description is probably most accurate for past sociotechnical experiments, and this approach will undoubtedly receive more widespread application in the future. Changes in the social and economic way of life are rendering traditional forms of work obsolete. Current problems such as poor quality workmanship, work stoppages, high absenteeism and turnover, and worker dissatisfaction are indications that the worker is not as productive or "fulfilled" as he could be. As organizations attempt to resolve these issues, there has been a growing realization that nothing less than a comprehensive approach to improvement in the work place will suffice. Sociotechnical research provides such a strategy. Moreover, its systemic orientation and democratic stance are in harmony with the thinking and values of a growing number of organizational members.

REFERENCES

Bregard, A., Golowsen, J., Hagen, F., Jolstad, E., Thorsrud, E., & Tyslano, T. *Novsk hydro: Experiment in the fertilizer factories*. Oslo, Norway: Work Research Institutes, 1968.

Coch, L., & French, J. Overcoming resistance to change. *Human Relations*, 1948, *1*, 512-532.

Cummings, T., & Griggs, W. Worker reaction to autonomous work groups:

Conditions for functioning, differential effects, and individual differences. *Organization and Administrative Science*, 1977, *7*, 87-100.

Cummings, T., & Molloy, E. *Strategies for improving productivity and the quality of work life*. New York: Praeger, 1977.

Cummings, T., Molloy, E., & Glen, R. A methodological critique of 58 selected work experiments. *Human Relations*, in press.

Cummings, T., Molloy, E., & Glen, R. Intervention strategies for improving productivity and the quality of work life. *Organizational Dynamics*, Summer 1975, *4*, 52-68.

Cummings, T., & Srivastva, S. Chaper 8: The estimating and die engineering experiment: A case study of white-collar department redesign. In T. Cummings & S. Srivastva, *Management of work: A socio-technical systems approach*. Kent, Ohio: Comparative Administration Research Institute, 1977a.

Cummings, T., & Srivastva, S. Chapter 9: The wheel-line experiment: A case study of blue-collar work design. In T. Cummings & S. Srivastva, *Management of work: A socio-technical systems approach*. Kent, Ohio: Comparative Administration Research Institute, 1977b.

Emery, F., & Thorsrud, E. *Form and content in industrial democracy*. London: Tavistock, 1969.

Emery, F., Thorsrud, E., & Lange, E. Field experiments at Christiana Spigerverk (Doc. T. 807). London: Tavistock, 1970.

Englestad, P. Socio-technical approaches to problems of process control. Paper presented at the Fourth Fundamental Research Symposium: Paper Making Systems and Their Control, Oxford University, 1970.

Foster, M. *Analytical model for socio-technical systems* (Doc. No. 7). London: Tavistock, 1967.

Gorman, L., & Molloy, E. Job restructuring in the ledger department of a bank. In L. Gorman & E. Molloy, *People, jobs, and organizations*. Dublin: Irish Productivity Centre, 1972.

Prestat, C. *Une experience de groupes semi-autonomes*. Paris: Foundation Internationale des Sciences Humaines, 1971.

R. G. Barry Corporation. *Small team production*. Columbus, Ohio: R. G. Barry Corporation, undated mimeograph.

Rice, A. K. Part III: The experimental reorganization of automatic weaving: The social reorganization of a technologically disturbed production system. In A. K. Rice, *Productivity and social organization: The Ahmedabad experiment*. London: Tavistock, 1958a.

Rice, A. K. Part IV: The experimental reorganization of nonautomatic weaving: The creation of a new socio-technical system. In A. K. Rice, *Productivity and social organization: The Ahmedabad experiment*. London: Tavistock, 1958b.

Trist, E., & Bamforth, K. Some social and psychological consequences of the longwall method of coal getting. *Human Relations*, 1951, *4*, 3-38.

Trist, E., Higgin, G., Murray, H., & Pollock, A. Chapter XIII: Face team organization and maintaining production. In E. Trist et al., *Organizational choice*. London: Tavistock, 1963a.

Trist, E., Higgin, G., Murray, H., & Pollock, A. Chapter XIV: Work load stress and cycle regulation. In E. Trist et al., *Organizational choice*. London: Tavistock, 1963b.

Van Gils, M. Job design and work organization in industrial democracy in the Netherlands. Personal communication from Netherlands, 1969.

Van Liet, A. *A work structuring experiment in television assembly* (T.E.O. Special). Eindhoven, Netherlands: Philips N.V., 1970.

Vossen, H. *Experiments in the special miniature lamp department of Philips N.V.* (Internal report). Eindhoven, Netherlands: Philips N.V., 1969.

Walton, R. How to counter alienation in the plant. *Harvard Business Review*, 1972, *12*, 70-81.

Socio-Technical Systems in Weaving, 1953-1970: A Follow-Up Study

Eric J. Miller

In 1953-1954, experimental changes in work organization were introduced in automatic and non-automatic loom-sheds of an Indian textile mill (Rice, 1953, 1955a, b, 1958, 1963). In each case semi-autonomous work-groups were established with responsibility for production and routine maintenance on a group of looms. Subsequently, these forms of working were extended to other loom-sheds. In 1970 a follow-up study of the 'group system' in four locations, including the two original sites, showed that in one–the non-automatic experimental loom-shed–the work organization and levels of performance had remained virtually unchanged over the 16 years, while in a newer automatic loom-shed group working had largely disappeared. Considerable regression had occurred in the site of the automatic loom experiment, and the remaining non-automatic site displayed variation in modes of group working. The paper suggests that the effective persistence of the 'group system' in at least one area implies that the assumptions in Rice's original experiment were substantially confirmed. Regression in the two automatic loom-sheds is explained in terms of failure to maintain the necessary boundary conditions for group working in the face of progressive changes in the market and other environmental factors. In conclusion it is suggested that Rice's innovations were more radical than was recognized at the time.

Reprinted from Eric J. Miller, "Socio-Technical Systems in Weaving, 1953-1970: A Follow-Up Study." HUMAN RELATIONS, 1975, *28*, 349-386. Used with permission of Plenum Publishing Corporation.

INTRODUCTION

The concept of a production system as a *socio-technical system* was introduced by Trist (Trist & Bamforth, 1951). It draws attention to the symbiosis of a technical system, consisting of equipment and a process lay-out, and a social system, through which the people performing the necessary tasks are related to one another. Each system has properties that are independent of the other, but they are also inter-dependent. Many studies have shown that attempts to maximize efficiency of the technical system, if social and psychological needs are not also satisfied, lead to sub-optimal performance; the converse is also true. This concept led to a productive search for 'the "right" organization that would satisfy both task and social needs', though somewhat later it has been recognized that 'there are settings where elegant solutions of this kind cannot be found or where, if found, they introduce new and intractable constraints' (Miller & Rice, 1967, p. xii).

The first explicit use of this concept in the experimental design of a production system was by Rice in 1953 at The Calico Mills, Ahmedabad, India.[1] '. . . Attempts were made to take into account both the independent and inter-dependent properties of the social, technological and economic dimensions of existing socio-technical systems, and to establish new systems in which all dimensions were more adequately interrelated than they had previously been' (Rice, 1958, p. 4). In three papers (1953, 1955a, b) and in the book from which that quotation is taken (*Productivity and social organization: the Ahmedabad experiment*) he describes an experimental reorganization of weaving, first in an automatic loom-shed and secondly on non-automatic looms. In both instances, internally led groups of workers were formed, each group being responsible for production and line maintenance of a group of looms.

The book also describes concomitant changes in management organization that the Company introduced while Rice was acting as consultant during the period 1953-56. The story of management reorganization up to 1961 is continued in Rice's second book, *The enterprise and its environment* (1963), which also gives up-dated information on the two loom-shed experiments.

---

[1]'The Calico Mills' is the name by which the Ahmedabad Manufacturing and Calico Printing Company is generally known. In the rest of this paper we refer to it simply as 'the Company'. This is to avoid confusion, since one of the spinning and weaving units within the Company's Textile Division carries the same name. This we call 'Calico Mill', or 'Calico'. The other spinning and weaving unit with which this paper is concerned is Jubilee Mill (or 'Jubilee').

This work of Rice's has been widely cited. For example, it is described by Katz & Kahn (1966) as 'by and large . . . an amazing success story' (p. 455). McGregor (1960, 1966) refers to it as a case of collaboration leading to integration; Myers (1959) as an example of involving workers in shop floor decision-making; Davis (1962, p. 497) as an example of a work situation promoting group cohesiveness; and Likert (1961, pp. 38-43) as illustrating how such cohesiveness can result in better performance figures.

For the most part, Rice's methodology has been either accepted or actively praised (for example, by Davis, 1967). Vroom & Maier (1962) are more cautious, saying that it is 'difficult to draw unequivocal conclusions concerning the underlying processes' (p. 438). One dissenting voice is that of Roy (1969), who criticizes the methodology and suggests that in his interpretation of the results he achieved, Rice had greatly underestimated the extra cash incentive offered to the group workers.

Rice was to have revisited the Company in his consultant role in 1969. Part of his brief was to examine the implications for work organization of further technological change in the textile industry. This visit had to be abandoned because of an illness which led to Rice's premature death towards the end of 1969. Accordingly, the author, who had during 1956-58 worked for the Company as an internal consultant in collaboration with Rice, went as a substitute for him to Ahmedabad for one month in July/August, 1970. This seemed to be a good opportunity to review the present state of the socio-technical system that Rice had helped to introduce, to analyse changes that had occurred and to try to refine the assumptions on which the original system was designed. Accordingly, permission was sought from the Company to deploy a second Tavistock staff member, A. F. Shaw, in a research capacity. The Social Science Research Council made a grant for this purpose.[2]

This paper, then, presents the principal findings from a possibly unique follow-up study of an experiment in work organization, the data spanning a total of 17 years. The remainder of the paper is divided into four parts:

I. The initial experiments and subsequent developments.
II. Research methods.

---

[2]The Social Science Research Council covered Shaw's travel to and from India and also salary costs; while in India Shaw stayed as a guest of the Company. I would like to thank both these bodies for making this piece of work possible. I am also grateful to Shaw for his preliminary analysis and interpretation of the data.

III. Findings.*
IV. Discussion.

THE INITIAL EXPERIMENTS AND SUBSEQUENT DEVELOPMENTS

Rice (1958) provides the principal base-line for the present study. Here I summarize only briefly the experimental changes carried out first in the automatic loom-shed at Jubilee Mill and second in the non-automatic loom-shed in Calico Mill.

The Jubilee Automatic Loom-Shed Experiment, 1953-54

New automatic looms had been recently installed. Prior to the experiment the weaving process had been broken down into component tasks and the number of workers allocated to each component had been determined by work studies. The resultant pattern was of an aggregate of individuals with confused task and role relationships and with no discernible internal group structure. Efficiency[3] was lower and damage higher than target figures. Productivity was, in fact, no better than with non-automatic looms.

Rice's assessment was that on automatic looms the title of 'weaver' for an occupational role was no longer appropriate: the weaver was now the loom, and all workers including the 'weavers' serviced machines. The tasks performed could be differentiated into two main types: those concerned with weaving, and those concerned with 'gating' (fitting new beams of warp-yarn) and loom maintenance. There were in addition only minor ancillary services. Rice then proposed the idea of a group of workers for a group of looms. The theoretical numbers required for blocks of 64 looms were calculated. Three natural grades within a work-group were found, instead of the nine grades existing. They were designated by letters only; rates slightly in excess of existing rates were fixed; and it was decided to pay piece rates to the whole group.

Shed supervisors and workers spontaneously took possession of the reorganization, the workers themselves immediately organizing four experimental internally led small groups. This was in March 1953. These groups abandoned the old titles, using only the new letter grades; and

[3]Efficiency is measured in terms of the number of 'picks' (weft-threads) inserted during a given period (usually a shift), expressed as a percentage of the number that would have been inserted if the loom had run continuously throughout that period.

*Editor's Note: "Part III. An Outline of Findings" is not included in this text.

after an immediate increase in efficiency at the cost of increased damage and inadequate maintenance, they settled down at a new level of performance in which efficiency was higher and damage lower than before reorganization.

When the form of organization was extended to the rest of the shed, and a third shift started, the efficiency was maintained for several months, but in October and November 1953 it dropped steeply through a period of five weeks. At the same time the figures for damage rose steadily. Investigation showed that each group had to contend with variations in the sort[4] woven; there had been insufficient spare workers; training of new and existing workers in the new methods of organization had been neglected and also diffused throughout all groups; and the basis on which the original experimental groups had been formed had not allowed sufficient time for group leaders to perform the task of leading. As a result of the difficulties caused by these factors, group members had regressed to earlier working habits more appropriate to individual than group working.

Various corrective measures were taken, including establishment of basic rates of pay, segregation of training from production, and implementing a policy of minimizing the variety of sorts in any one group. Recovery was rapid.

> 'It was concluded that the first spontaneous acceptance of the new system and the subsequent determination to make it work were due primarily to the workers' intuitive acceptance of it as one which would provide them with the security and protection of small group membership which they had lost by leaving their villages and their families to enter industry. At the same time the new system allowed them to perform their primary task effectively and thus provided them with an important source of satisfaction' (Rice, 1958, p. 110).

The Calico Non-Automatic Loom-Shed Experiment, 1954

The presenting problem was to improve the quality of cloth woven on non-automatic Lancashire looms. The 560 looms in Calico A Loom-shed were operated on two shifts. Low output and high damage rendered three-shift working uneconomic. Studies of the existing production system showed that weavers, who each ran two looms, were, in their formal role-relationships, independent isolates, but informally made

[4]'Sorts' or types of cloth are classified by the 'count' (thickness) of yarn and the numbers of warp and weft threads to the inch.

mutually helpful relationships with each other. They were subject to the authorities of three departments and frequently had to visit other parts of the mill to obtain weft yarn and to deliver woven cloth. They nevertheless performed an integrated 'whole' task, the conversion of yarn to cloth, and were responsible for the quality and quantity of cloth woven.

An experimental shed was set up, in which small work-groups were constituted to perform all the tasks of weaving and maintenance on groups of forty looms, all of which wove the same kind of cloth continuously. Four natural grades within a work-group were found: group leader, front loom worker, back loom worker, and helper. These group members performed interdependent tasks and formed interdependent relationships. An aggregate of 22 workers in the conventional system was thus replaced by an internally structured group of 11. Minimum basic rates were fixed higher than had previously been paid. Bonus was paid both for quantity and for quality. The bonus was paid, at the workers' own request, for group rather than individual performance.

The experimental period lasted 10 months (January-November 1954) during which both management and workers progressively 'tested out' each other's sincerity and willingness to co-operate. Gradually permissive and collaborative relationships based on mutual trust were built up. The building of such relationships between management and workers and the stabilizing of intra-group relationships were assisted by the institution of informal group meetings and of more formal conferences. In the conferences the whole executive chain responsible for production was present and at the workers' request all shifts attended the same conferences, some coming in, in their own time, to attend.

After four months, a campaign against 'rationalization' flared up in the city. Employees in the experimental shed were identified as supporters of rationalization: the mill gates were picketed in an attempt to stop them working; their families were threatened; they were hooted at as they walked through the mill and one worker was stoned. More members of the experimental shed than formerly exercised their right to return to their former jobs during this time, but there were always more applicants for work in the shed than there were vacancies.

After a very slow start the efficiency of the shed climbed steadily, with a setback during the height of the campaign against rationalization, until during the final phase of the experimental period it was somewhat higher than equivalent efficiencies in other loom-sheds. The quality of cloth, after being worse than that woven in other sheds for the first phase, steadily improved, again with a setback during the anti-rationalization campaign, and finally settled down at a better standard than was achieved in other sheds.

Improvement in quality was achieved partly by a reduction in loom speeds of 11 per cent. Despite efficiency increases, therefore, output per loom hour was reduced. On the other hand, labour productivity increased substantially and this, in conjunction with the higher average price per yard obtainable from the improved cloth, made three-shift working viable. Compared with the existing two-shift two-loom system, the experimental three-shift group system showed the following results: the mean earnings of the group were 55 per cent higher; the cost per loom was 13 per cent higher; the output was 21 per cent higher; and the number of damages 59 per cent less. These results were achieved on the most difficult sort regularly woven in the mills.

In 1955 the group method of working was introduced on the third shift in a non-automatic loom-shed at Jubilee Mill, where, on simpler sorts, similar earnings were achieved, with a 4 per cent rise in output and a 47 per cent reduction in cost. Rice concluded:

> 'The immediate practical result of the experiment has been to demonstrate that in the Calico Mills the breakdown of the "whole" task of weaving into component operations, each performed by a different worker, and the re-integration of the workers into an internally structured work-group that performs the "whole" task on a group of looms, can be accomplished in one process provided that permissive and collaborative relationships can be built up between all those concerned . . .
>
> 'The experimental system has established new norms of performance and earnings for non-automatic weaving. The conclusion was reached that the acceptance of the new system and the determination to make it work were due to its providing more opportunities for effective task performance and for the building of stable and secure small work-group relationships than those existing in the conventional system with which the traditional norms of performance and wages were associated' (Rice, 1958, p. 166).

Subsequent Developments

Graphs showing follow-up performance figures in both experiments during the period 1954-60 are given by Rice in his second book (1963, pp. 111-112). These show that the levels of efficiency reached during the experimental phases were largely sustained and in the case of the non-automatic looms even slightly improved during the subsequent six years. Similarly, the levels of damage went down and stayed down. Whatever argument there might be with the explanations that Rice proffered, these data show that new norms of performance were set and successfully maintained.[5]

Various extensions of group working occurred in the period up to 1970. To take non-automatic looms first, Calico A Loom-shed included the 120 looms in the Experimental Shed plus 440 in the so-called Pit Shed. The latter were also transferred to group working in 1956-57, with improved performance. On the other hand, Calico B Loom-shed contained 660 non-automatic looms weaving a wide variety of often complex sorts and these were therefore technologically unsuitable for a method of group working. In this shed the pre-1953 method of working, which involved one weaver looking after two looms and a jobber providing maintenance services for a larger number of looms, remained in operation. At Jubilee the 452 non-automatic looms continued to operate under a group system on the third shift only.

On automatic looms, group working became universal. At Jubilee Mill, 48 of the 288 automatic looms on which the original experiments had been carried out in 1953 were disposed of in the late 1960's and an additional 128 looms of a different type, and not new, were added. Also during the 1960's, the Company was able to purchase additional automatic looms and a new building, the Auto Shed, was constructed at Calico Mill. In 1970, this contained 656 looms. The group method of working which had been successful on the original Jubilee automatic looms was applied to the new Calico Auto Shed, but management had been disappointed initially by low productivity. It took several years for efficiencies to climb and damages to fall to the levels hoped for. And

---

[5]Comparisons of performance data need to be treated with a good deal of caution. It is virtually impossible to make precise comparisons between different types of looms weaving different types of cloth. 'Before and after' comparisons on the same looms weaving a similar type of cloth are more reliable. Apart from minor changes which may be made in the methods of calculating efficiency and in inspection standards, however, a reduction in loom speeds can lead to improved efficiency figures but a decline in output. Speeds were adjusted during the first experiments but so far as could be checked, not subsequently. The performance data therefore probably do show real trends. It has also to be noted that although fluctuations in the quality of yarn can increase or decrease the work-load on the looms, in the form of organization designed by Rice provision was made to increase or decrease the number of workers according to sample measurements of the number of breaks occurring in the warp and weft yarn. Thus, although labour productivity would change, the broad levels of efficiency and damage should remain substantially the same. Maintenance of the looms, which also affects both efficiency and damage, is the responsibility of the group in this type of work organization. There are, however, limitations on the extent to which the work group can offset the consequences for efficiency and damage of poor quality of supplies (such as shuttles) and spare parts.

even at the end of this period the damage rate was still relatively high by good international standards. However, in view of problems of comparison, it is difficult to determine how much of this difference was attributable to the performance of the workers on the looms themselves, how much to the quality of the yarn and supplies available, and how much to changes in criteria of measurement.

Other Trends, 1955-1970

During this period, progressive changes were occurring in the Company's markets and sources of supplies.

Since the early 1950's, it had been the avowed policy of the management of the Company to raise norms of productivity and quality, and Rice's experiments themselves were designed as means to that end. The effects of this policy were reflected in the Company's record of profitability, which was surpassed by few other mills in India (Rice, 1963, p. 108). As the Company enhanced its reputation for design and quality in the Indian markets and correspondingly was able to command a progressively higher average price per yard of cloth sold, so its management aspired to becoming increasingly competitive in export markets, not only within the relatively less developed countries of the Far East, the Middle East and Africa, but also in the more stringent markets of the West—for example, shirting for the mass-production garment industry, which demands continuous pieces of flawless cloth.

There were two significant constraints. First, there was a limited availability of long staple cottons required for the fine and most profitable fabrics, since these had to be imported, mainly from Egypt, and India has maintained a tight control over imports for many years. To maximize use of this raw material, high standards were demanded in spinning and weaving. Second, Governmental regulations, imposed to protect the hand-loom industry, prevented the installation of additional power-looms. Consequently, there was little scope for increasing profits through expansion and greater emphasis on productivity and profit per loom. Top management pressures to maintain and improve quality and output were thus strong and persistent.

Import controls also restricted availability of loom spare parts and supplies, such as shuttles. Domestically manufactured substitutes were often of low quality. This led to damage and reduced efficiency and to some deterioration in the general condition of the looms.

RESEARCH METHODS

My consultancy visit to the Company in July/August 1970 produced a good deal of data about technical and organizational changes. I interviewed many managers, not exclusively in weaving, and also had several extended discussions with weaving work-groups, whose members spoke fairly freely about their current work experience. More detailed data was obtained by Shaw during a three-month field visit between September and December 1970. Three main methods were used: observation, interviews, and examination of performance and other records.

Intensive observation in the weaving sheds provided a means of comparing methods of working actually in use with historical and contemporary statements about the methods.

A total of 78 workers were interviewed, according to the schedule shown in Table 1. They fell into seven samples, covering four areas, chosen so that comparisons could be made between the different technical systems and also between those workers who had been in the company before 1953, when the group system was originated, and those who had joined subsequently. Selection was made from a limited number of work-groups in each department, avoiding any groups temporarily experiencing unusual working conditions, and within the chosen groups selection was not entirely random in that an effort was made to ensure that no category of workers was unrepresented.

Since all but two of the interviews had to be conducted in either Gujarati or Hindi, an employee of the Company (but from an area remote from production) was trained as an interviewer. He was briefed in advance about areas to be explored, and at the end of each interview he was de-briefed in detail by the fieldworker, nuances and idioms being examined and the interview being looked at as a whole rather

Table 1. Samples of Manual Operatives

Mill	*Department/ Type of loom*	*Type of work organization*	*Date of joining Co.*	*Number*
Calico	'A' Loom-shed non-automatic	Group system	Before 1953 1953 and later	12 12
	'B' Loom-shed non-automatic	Non-group	Before 1953 1953 and later	12 12
	Automatic Loom-shed	Group system	—	12
Jubilee	Automatic Loom-shed	Group system	Before 1953 1953 and later	9 9

than as a collection of sentences. In this way, some of the subtlety of the information lost through an interpreter was recovered.

Managers and supervisors from these same four units were interviewed both singly and in groups.[6] The meetings were relatively unstructured and often prolonged; some were continued at intervals over several days or even weeks. With some supervisors the interpreter was used, but most respondents spoke fluent English.

The records available included field notes prepared by Rice mainly between 1953 and 1958, and by myself between 1956 and 1958 (when I was employed by the Company as an internal consultant) and in 1970. These allowed some comparisons to be made between current and past practices, though virtually all the 'baseline' data were in fact taken from Rice's published material. We also had access to detailed production data. These included monthly average efficiency and damage for each group on each shift for the period January 1969 to October 1970. However, the difficulties of interpreting such data unambiguously have already been mentioned.

There was one defect in the research design. Calico A Loom-shed (non-automatic) comprised both the Experimental Shed and the Pit Shed, but because these had been weaving the same sorts of cloth on the same types of looms and using ostensibly the same work organization since 1956, we had treated them as a single sample. Our research, however, revealed significant differences, and subsequently some of the data has been re-analysed to distinguish between the two sheds, here called 'A (Exp)' and 'A (Pit)'.

Two aspects of the circumstances of the fieldwork need to be recorded here, in that they might be held to have affected the data.

The first concerns managerial reactions to the research. The Company's previous experience of the Tavistock Institute had been in the client/consultant relationship. Rice's publications were a by-product of his consultancy. The Company Chairman and General Manager of the Textile Division had agreed to the study, though it was mutually accepted that specific sanction had to be obtained from each superior before anyone in his chain of command could be approached for interview. In practice it was found quite difficult to start fieldwork. Contact

---

[6]In their internal structure, the units were similar. On each shift, a shift supervisor had responsibility for a number of groups. One or more shift supervisors per shift reported to a section head responsible for a particular set or type of looms on all three shifts. Section heads in turn normally reported to a weaving master, responsible for an overall loom-shed, e.g. Calico A, Jubilee Auto. The weaving master in charge of Calico Auto had no responsibility for the preparation of the beams and pirns used on his looms; the other two weaving masters did have such responsibility.

with manual operatives was continually held up on a variety of pretexts, which suggests that there was a good deal of underlying suspicion about the project—both fear about the supposed reasons for the study and scepticism about its worth. The fear and scepticism diminished as the fieldworker made it clear that he respected managers' concerns and that his research might be helpful for them as well as to himself. As will be seen, one factor in the resistance was a wish to suppress a discrepancy between a form of work organization actually operating and what top management believed to be operating.

The second circumstance relates to labour relations. During the study the union—the Textile Labour Association—called a strike of its members in the city of Ahmedabad as a whole. Historically, the workers in this Company had tended not to take part with those from other mills in city-wide strikes of this kind. In this case some did. The industrial climate did not seem to diminish workers' willingness to be interviewed. Although attendance was voluntary, no difficulty was experienced in obtaining respondents and several workers who had not been asked to attend an interview came to express views about working arrangements in the mills. Our impression is that the circumstances made respondents less inhibited about expressing their views than they might otherwise have been, given the prevailing climate of overt compliance to authority.

DISCUSSION

Between 1953 and 1956, Rice helped to introduce and develop two types of 'group system' in automatic and non-automatic weaving. There were certain differences between these two systems related to technological differences between the two types of looms, but the underlying concept of work organization was the same for both: a small, internally led group of workers, responsible for the whole task of weaving on a group of looms.

By 1970, the so-called 'group system' encompassed widely different methods of working.[7] The main common thread was that the individual's pay was to some extent affected by efficiency and damage on a group of looms.

The group system in Calico A (Exp) in 1970 differed little from Rice's description of it some 15 years before. Group identification was high; members of the group co-operated with one another in their work; the

[7]It is worth noting that the term 'group system' has been taken over as an English word amongst Gujarati and Hindi-speaking people in the mills; there is no generalized concept of 'group' in these languages.

group leader exercised a boundary function; and supervisors seldom intervened with individual group members. It is notable that this was still in 1970 called the experimental shed, which emphasizes the strong link to the past. Slightly more than half the workers belonged to the original experimental groups and associated what they felt to be their present high status with their historical achievements at the time of the experiment. The sort of cloth woven was still the same, and, as we have seen, norms of performance seemed to have persisted for 14 years. It was as though the shed had been held within a kind of stasis—a monument to the original experiment.

In A (Pit), the group system in 1970 resembled that of A (Exp), though performance was significantly inferior and there was greater variation between groups in methods of working. Certainly in some there was more of a differentiation of function between group members with correspondingly less co-operation. Compared with A (Exp) there was somewhat less group identification and a tendency for the supervisors to take over boundary control functions that in the experimental shed belonged to the group leader.

In Jubilee Auto the shift in this direction was even more pronounced. Workers in many groups were held responsible for operating a specific sub-set of looms; there was little mutual help; identification with the group was limited; the group leader was preoccupied with his loom maintenance activities; and to a considerable extent the shift supervisor directly controlled the activities of individual workers. In this respect it is interesting to look back in more detail at Rice's account of the regression that occurred in the experimental groups at Jubilee towards the end of 1953 after they had been operating for some six months (1958, p. 94). At that time there was a reversion to individual work habits with individuals taking responsibility for specific tasks on a specific number of looms, even though this meant that some of them were underloaded while others were overloaded. There was little or no mutual help. 'Faced with the lack of group leadership and the regression to earlier work habits . . . the supervisors themselves tended to regress to earlier patterns of management behaviour. They by-passed the group leaders and dealt directly with workers. . . . The intervention by the supervisors in the organizing of group tasks tended further to destroy the internally structured leadership of the groups' (pp. 94-95). Several factors had conspired to produce this regression at that time. The main factor was that a number of workers had been appointed to positions in the group for which they were inadequately trained and skilled, but in addition there was 'the constant confusion due to mixed and experimental sorts, the shortage of spare workers, the increased work caused by . . .

greater absenteeism . . . and the disturbance of building and loom erection' (p. 95). The 1953 regression had been followed by an investigation and an introduction of remedial changes over the next few months, with the result that the groups regained and exceeded the higher norms of performance that had been attained during the initial experimental period. Group functioning in 1970 differed in one additional respect from the regressed phase of 1953. Whereas in 1953 the group leaders had tended to pay attention to production, thus reverting to the role of worker and neglecting the leading of their groups, in 1970 the group leaders were almost wholly preoccupied with maintenance.

Finally, in Calico Auto, except for the brief phase when the charade was put on for the benefit of the research worker, the method of working could scarcely be described as a group system at all. Group identification and internal co-operation were virtually lacking. The terminology to describe different categories of workers was much the same as in other mills. The group leader was effectively a jobber with maintenance responsibilities, though still somewhat resentful of the fact that supervisors were taking responsibility for internal management of the group and intervening with individual group members.

We therefore have to try to account for the persistence in Calico A (Exp) of the socio-technical system developed by Rice, for the emergence of a discrepant system in Calico Auto, and for the development in Calico A (Pit) and in Jubilee Auto of systems appearing to be intermediate between these two extremes.

I have described two developments in the relation of the Company to its environment that impinged on the loom-sheds. First, the efforts to enter more lucrative markets led to pressure to maintain and improve outputs and quality; and second, inability to procure spare parts and supplies of the appropriate standard increased both the maintenance work-load and the need for attention in weaving activities. The latter factor may have exacerbated the tendency of production incentive schemes in general, especially with three-shift working, to maximize production in the short term at the expense of maintenance of machines. Probably, therefore, the group leaders in 1970 were facing an accumulated legacy of sub-optimum loom maintenance.

Now the method of working had been designed so that each loom-group, as a socio-technical system, had a certain amount of resilience to absorb and adjust to variations in its inputs without invoking regulatory interventions from outside. One feature was that the size of the weaving sub-group could be increased if yarn inputs were sub-standard. Apart from that, the main source of resilience was flexibility in task allocation, in place of a rigid division of labour between categories of workers and between individuals. Rice went some way towards specifying the boundary conditions within which the two systems could be expected to maintain a steady state: for this the regression of late-1953 in Jubilee Auto provided illuminating data. The resilience of the groups was evidenced in Jubilee Auto by the fact that they were able to weave varieties of cloth that had previously been found 'unweavable' on the looms concerned; and in Calico A (Exp) by the workers' persistence with the experimental group system, though with some setback in performance standards, in the face of picketing, threats and even physical assault by some other workers during the 'anti-rationalization' campaign of November 1954. These experiences confirmed Rice's proposition that

> 'the performance of the primary task is supported by powerful social and psychological forces which ensure that a considerable capacity for co-operation is evoked among the members of the organization created to perform it' (Rice, 1958, p. 33),

and suggested that the socio-technical systems that had evolved optimized task and 'sentient' needs (cf. Miller & Rice, 1967, pp. xxii ff. *et passim*).

Rice's discussion of the regression in Jubilee implies that a reversion to a more rigid differentiation of labour could be taken as a symptom that the group as a social system had exceeded the limits of its capacity to accommodate to external change. When such symptoms are observed, the sophisticated managerial response is to seek ways of (a) reducing the sources of disturbance in terms of variability of inputs, and (b) increasing the resilience of the group—in other words, to adjust the technical and/or social system in such a way that optimization between them is restored. The corrective measures taken during the first half of 1954 in Jubilee Auto had these characteristics. The boundaries of the groups were restored and they recovered their viability and resilience. The kinds of decisions required to achieve this, however, had in some cases to be made at a fairly high managerial level. This was possible while group methods of working were still in an experimental phase and while, correspondingly, senior managers were very directly concerned in monitoring the effectiveness of the experiment. Subsequently, as group working 'settled down' as an established and less controversial form of work organization, top management involvement was gradually withdrawn, and indeed within two or three years the Chairman was saying that he scarcely ever visited the experimental shed because he knew that everything there was going all right. The reorganization of management undertaken in the Company between 1954 and 1960 with Rice's help (see Rice, 1958 and 1963) involved the drawing of organizational boundaries round not merely these primary work-groups but also a series of progressively wider systems—the shift, the section, the loom-shed, the mill—each of which was designed to have corresponding resilience. The role of a supervisor or manager was conceived in terms of regulating the boundary of the system for which he was responsible so as to maintain its internal resilience in the face of change; and the corollary of this of course was that he was expected to draw the attention of his superior to sources of disturbance to which it was beyond the capacity of his own system to adjust. The position in the loom-sheds in 1970 suggests that these successive boundaries had not been effectively controlled. Disturbances had been transmitted to and into the work-groups themselves. The concept of boundary control, even if understood theoretically, had not been implemented in practice and, what is more, it had evidently not been considered important to ensure that supervisors understood its implications for their own jobs in relation to the groups. Consequently, by their direct internal interventions supervisors further helped to destroy the resilience of the groups and to foster the more rigid differentiation of individual tasks.

One factor of possible relevance here was the progressively diminishing 'bite' of the bonus system. The group worker's wage packet included three components—a rate for the category, a dearness allowance and a bonus calculated on the first rate. Between 1953 and 1970 basic rates had increased only marginally while the dearness allowance component had trebled. Thus, for example, for a B worker in A (Exp) in 1954, with a basic rate of 115 rupees per month and a dearness allowance of 62 rupees, his average bonus of 28 per cent on his basic rate was 18 per cent on total earnings. By 1970, when the figures were 125 and 180 rupees respectively, the 28 per cent would have yielded a net bonus of only 11½ per cent. This is an example of a factor that might have reduced — or more importantly have been held by supervisory staff to reduce—the capacity of groups to maintain their self-regulating capabilities. (It may also be noted that maintenance of high levels of performance in A (Exp), in spite of the fact that a reduction of effort and thus of efficiency would have led to only a marginal reduction of earnings, indicated that the effects of the cash incentive may have been smaller than Roy (1969) suggests.)

Therefore, between 1953 and 1970 the groups were faced with a constant series of minor readjustments. Some of these adjustments were recurring, since seasonal or fashion variables required cyclical changes from one sort of cloth or one pattern of cloth to another. However, the loom-sheds were also having to adjust to the progressive attempts of the Company's management to improve its overall position *vis-à-vis* a changing set of competitors. Therefore, it was not simply a question of maintaining an equilibrium within a broadly static range of possible steady states but of attaining an equilibrium within a constantly altering trajectory.

We still have to explain the differences between loom-sheds observed in 1970. In fact the nature of the pressures was such that they were likely to be experienced maximally in Calico Auto and minimally in A (Exp). Calico Auto was weaving the finest and most expensive sorts with a high fashion component (thus implying the need for cyclical adjustment) and these were also the types of cloth most influenced by the Company's search for new and profitable markets. It was in this shed that marginal improvements in production and quality would produce the greatest pay-off in terms of improved profits. In A (Exp) on the other hand, although the type of cloth woven was by no means simple—in fact there had been a deliberate decision when the experimental shed was set up to select the most difficult sort commonly woven—it was a product for which there was a reliable market and in which few if any improvements in quality standards were required since the acceptably higher norms had been attained at the end of the experimental period in 1954. Here it had been possible to maintain a steady state. Jubilee Auto was in an intermediate position, in that the fabrics woven were neither so fine nor so profitable as in Calico Auto, though at the same time there had been continuing pressure over the years to move to marginally more profitable and correspondingly more difficult sorts and to improve quality standards. In A (Pit) the evidence is less clear. It seems likely, however, that the process of extending the group system to this shed may have involved too inflexible a transfer of a group structure, without giving the groups enough room or help to discover their own modes of working within the appropriate culture. Some found their own resilient modes; but in other instances they failed to develop adequate boundaries within which self-regulation could occur, so that supervisory intervention became integral to the regulatory function within groups.

Problems of spares and supplies impinged differentially in the same directions. Closer tolerances are required for automatic than for non-automatic looms and for finer than for coarser cloth.

Probably a further factor was the surviving proportion of original group members in each of the sheds. To the extent that group members had internalized this method of working they could be expected to be better able to maintain resilience and to resist incoming disturbances. As we have seen, more than half the workers in A (Exp) come into this category, a third in Jubilee Auto and a negligible number in Calico Auto. Thus, direct intervention by supervisors with individual workers which would be regarded as abnormal in, for example, A (Exp) would be accepted as normal and natural by those who had experienced more conventional modes of work organization and supervision elsewhere.

This also suggests that the need for individual recognition by one's superiors, which is quite strong in the Indian culture, was not really provided for within Rice's concept of group organization. The individual was to derive satisfaction primarily from the respect of his colleagues within the group. It would seem that this need could remain submerged so long as the group as a whole could be perceived externally as being sufficiently successful—though our observations in 1970 suggest that the role of the group leader, even in A (Exp), had been made more into that of a superior or boss and less of a *primus inter pares* than Rice had originally envisaged. One would therefore postulate that diminishing experience of success, accompanied by a withdrawal of group leaders from the leadership aspects of their roles into a greater preoccupation with maintenance, would lead to the reactivation of the need for recognition from outside the group. This would increase individual workers' readiness to accept direct supervisory intervention from outside and correspondingly move the mode of intra-group relations away from co-operation in the direction of competition.

In conclusion, it may be suggested that Rice's innovations were more radical than was recognized at the time. Although Myers (1959) noted that the changes reflected a democratic, participative philosophy which was at variance with the paternalistic, authoritarian philosophy generally prevalent in Indian management, and Rice himself stressed the importance of developing 'permissive and collaborative relationships' which brought workers, supervisors and managers together in problem-solving approaches, both these comments are more apposite to the process of introducing the experimental changes than to the nature of the ongoing socio-technical systems that emerged. The permissive element in these systems was their built-in capability for self-regulation, which made them resilient over time within certain boundaries of stability.

Rice's application of systemic concepts to organization had been much influenced by the biologist, von Bertalanffy, who introduced the idea of a quasi-stationary equilibrium in an open system (1950). Understanding of ecological systems has developed considerably in the last 20 years and it is now well-known that intervention in such systems needs to be circumspect if unanticipated side-effects are to be avoided. Thus direct attempts to raise crop yields by applying pesticides and/or fertilizers will probably produce short-term benefits, but consequential changes in the wider ecological system are problematic and may in the longer term cancel any gains and even lead to regression to lower productivity than had prevailed in the first instance. Loss of variability and a contraction of the boundaries of stability mean a loss in the system's capacity for self-regulation. The goal of maximizing productivity in ecological systems has therefore become suspect—unless a very long time-scale is projected. Indeed, modern ecologists are suggesting that the appropriate 'conceptual framework for man's intervention into ecological systems . . . changes the emphasis from maximizing the probability of success to minimizing the chance of disaster' (Holling & Goldberg, 1971, p. 226); and nowadays some planners are learning the lesson from this, as these authors and others (for example, Friend & Jessop, 1969) have indicated.

The analogy with industrial production systems is suggestive, even compelling. Single-minded pursuit of efficiency goals nevertheless dies hard in industry.[8] Rice's approach, insofar as he was building a new resilience into the experimental socio-technical systems, was entirely consistent with the goal of 'minimizing the chance of disaster'; but in the prevailing industrial ethos it would have been difficult for him to put it in that way and still more difficult for others to perceive it. Paradoxically, because the prior pursuit of efficiency goals had resulted in a relatively unproductive work organization, Rice's efforts led to improved performance in the relatively short term and could thus be interpreted as being in harmony with efficiency goals—for example, Likert saw Rice's experiments as confirming that group cohesiveness leads to higher performance (Likert, 1961). It is much more plausible to suggest that the effect on performance of introducing a socio-technical system designed for long-term viability will depend on whether pre-existing levels of performance were high or low. Similarly, it is postulated that the performance of a system designed for minimizing disaster can almost always be improved by intervention designed to maximize efficiency: what is problematic is the length of time over which the improvement can be sustained. The most probable outcomes are a decline in performance, a multiplication of regulatory interventions, or both. If this is so, then the most remarkable outcome of Rice's experiments is that the 'group system' survived so completely in A (Exp) and, albeit to a lesser extent, in Jubilee Auto, during a period of considerable change and in a managerial environment in which efficiency goals largely prevailed. This suggests that the assumptions on which Rice worked have been largely substantiated. It also suggests that, quite independently of any notions of 'industrial democracy', the goal of designing systems to minimize the chances of disaster may be more appropriate to industrial organization than is generally recognized.

---

[8]Here I am using 'efficiency' in its more ordinary sense.

REFERENCES

BERTALANFFY, LUDWIG VON (1950). The theory of open systems in physics and biology. *Science, 3*, pp. 23-9.

DAVIS, K. (1962). *Human relations at work*. 2nd ed. New York: McGraw-Hill.

DAVIS, L. E. (1967). Job design and productivity. In *Studies in personnel and industrial psychology*, E. A. Fleishman (ed.). Homewood, Ill.: Dorsey, pp. 308-315.

FRIEND, J. K. and JESSOP, W. N. (1969). *Local government and strategic choice*. London: Tavistock.

HOLLING, C. S. and GOLDBERG, M. A. (1971). Ecology and planning. *Journal of the American Institute of Planners, 37*, pp. 221-230.

KATZ, D. and KAHN, R. L. (1966). *The social psychology of organisations*. New York: Wiley.

LIKERT, R. (1961). *New patterns of management*. New York: McGraw-Hill.

McGREGOR, D. (1960). *The human side of enterprise*. New York: McGraw-Hill.

McGREGOR, D. (1966). Why not exploit behavioural science? In *Leadership and motivation*, W. G. Bennis and E. H. Schein (eds.). Cambridge, Mass.: M.I.T. Press, pp. 239-275.

MILLER, E. J. and RICE, A. K. (1967). *Systems of organisation*. London: Tavistock.

MYERS, C. A. (1959). Management in India. In *Management in the industrial world*, F. Harbison and C. A. Myers (eds.). New York: McGraw-Hill, pp. 137-153.

RICE, A. K. (1953). Productivity and social organisation in an Indian weaving shed: an examination of the socio-technical system of an experimental automatic loomshed. *Hum. Relat., VI*, No. 4, pp. 297-329.

RICE, A. K. (1955a). The experimental reorganisation of non-automatic weaving in an Indian mill: a further study of productivity and social organisation. *Hum. Relat., VIII*, No. 3, pp. 199-249.

RICE, A. K. (1955b). Productivity and social organisation in an Indian weaving mill, II. A follow-up study of the experimental reorganisation of automatic weaving. *Hum. Relat., VIII*, No. 4, pp. 399-428.

RICE, A. K. (1958). *Productivity and social organisation: the Ahmedabad experiment*. London: Tavistock.

RICE, A. K. (1963). *The enterprise and its environment*. London: Tavistock.

ROY, S. K. (1969). A re-examination of the methodology of A. K. Rice's Indian textile mill work reorganisation. *Indian Journal of Industrial Relations, 5*, pp. 170-191.

TRIST, E. L. and BAMFORTH, K. W. (1951). Some social and psychological consequences of the longwall method of coal-getting. *Hum. Relat., 4*, pp. 3-38.

VROOM, V. H. and MAIER, N. R. F. (1961). Industrial social psychology. *Annual Review of Psychology, 12*, pp. 413-446.

The Comparative Impacts of Sociotechnical System, Job-Redesign, and Survey-Feedback Interventions

William A. Pasmore

From time to time, most organizations have an opportunity to change the ways in which employees perform organizational tasks. At times, these opportunities are welcomed as signs of progress; new facilities are constructed or older ones are revamped. At other times, opportunities for change come up in the form of problems—strikes, mechanical breakdowns, and layoffs. Whatever the causes of organizational change, managers turn to the behavioral sciences for suggestions and solutions. Unfortunately, what many find are not single, widely accepted solutions to their problems but a myriad of techniques, all of which supposedly provide answers to organizational concerns. It is not surprising, therefore, that managers opt for the technique that is the current fad or is best advertised. To date, only a handful of studies have compared the effectiveness of different types of intervention techniques, and their findings have not always been clear.

Without additional comparative research, even recommendations by experts about which techniques a manager should use tend to be less than scientific, if not heavily colored by personal bias. Consequently, some managers report success and others may not; the result is that the behavioral sciences are seen as unreliable for providing organizational assistance.

The aim of this paper is to help managers and experts alike make more informed choices about what intervention techniques to apply. The effectiveness of sociotechnical system, job-redesign, and survey-feedback methods will be studied in comparable units of a single organization. Both attitudinal and productivity measures will be used to assess the comparative effectiveness of these methods. It will be evident that in terms of improving employee attitudes, the technique chosen makes little difference; in terms of improving productivity, however, the method used appears critical.

A BRIEF DESCRIPTION OF THE INTERVENTION TECHNIQUES

Friedlander and Brown (1974) have classified interventions into two categories: human process and technostructural. The survey-feedback technique is representative of the former type; job-redesign and sociotechnical system interventions fall into the latter classification.

Human-process interventions "value human fulfillment highly and expect improved organizational performance to follow on improved human functioning and processes" (Friedlander & Brown, 1974, p. 325). The survey-feedback technique begins with a survey of employee attitudes and organizational characteristics. This information is tabulated and shared by each supervisor with his or her employees, who are asked to make suggestions for improvement; whenever possible, the suggestions are implemented. Although employees may suggest changes in technology or work content, we have found that instead they usually suggest changes in work methods or conditions *given their present jobs*, degree of responsibility, and so forth. This distinction between human-process and technostructural approaches to organizational change is important.

Technostructural interventions, in contrast to human-process interventions, "affect work content and method and affect the sets of relationships among workers" (Friedlander & Brown, 1974, p. 320). Technostructural techniques are based on the premise that increased employee responsibility and improved work methods will lead to greater satisfaction and higher performance.

Although both job-redesign and sociotechnical system interventions are technostructural in nature, job redesign is aimed at motivating individual employees by adding more variety, autonomy, responsibility, and feedback to their jobs. Generally, job-redesign interventions do not affect the technology used to produce outputs. Sociotechnical system interventions, on the other hand, have traditionally been focused on tasks performed by groups of employees and may actually alter the technology to provide better social relationships and production methods. These distinctions are important because, to the extent that levels of output are constrained by technological or group forces, changes in these variables may be required to bring about significant improvements in productivity.

Reviews of the literature by Friedlander and Brown (1974), Kahn (1974), and Srivastva et al. (1975) indicate both human-process and technostructural techniques have been fairly successful in improving employee attitudes. The reviewers concluded that technostructural interventions were more often associated with improvements in quality and productivity. In more detailed comparisons of the technostructural methods, the reviewers noted that job-redesign interventions were often associated with improvements in quality but not productivity, and sociotechnical system interventions were associated with improvements in both. Based on these reviews of the literature, the following hypothesis could be posited:

H_1: *Employee attitudes will improve following either human-process or technostructural interventions; however, productivity will improve only after sociotechnical intervention.*

The Study

To provide comparative data to test this hypothesis, in 1974 the author became part of a research team commissioned to evaluate the start-up of a new unit at a Midwestern production facility of a large food processing corporation. The facility employed approximately two hundred unionized hourly workers in two separate plants (Units "I" and "II"). The units were devoted to the processing of different products, although both were controlled by the same site-management group. The technologies employed in both units were highly sophisticated, requiring high levels of skill on the part of many operators. The work force consisted of nearly equal numbers of men and women.

The research team was commissioned after a decision was made by the home office of the corporation to revamp a major portion of the production facilities in Unit I due to changing market demands. Although the original employees would remain after the redesign, their work conditions would be quite different. The corporation decided to utilize sociotechnical system design principles in reconstructing the unit due to the success of such interventions elsewhere in the corporation. In addition to the substantial reconstruction of Unit I, a new first-level management group was hired to supervise its operation. Although the top management group and the supervisors of Unit II remained the same, it was considered necessary to hire and train new managers for the Unit I operation to support the increased employee responsibility called for by the design. Although traditional management techniques had been successful at the site, management had stressed technology concerns and production control and had a fairly hard-nosed approach to labor relations.

Naturally, there was some initial friction between the old and new management groups. The existing Unit I management group suspected that the corporation and other employees would pressure them to become less traditional in their approach to management and more like the new supervisors. The new supervisors, on the other hand, felt constrained by existing labor contracts and informal agreements. The site manager recognized these concerns and commissioned the research team to evaluate the start-up and the impact it had on the operation of the total site. He wanted the researchers to use whatever data they collected to improve the start-up and the functioning of the total site. This led the researchers to use survey-feedback techniques, which could provide data about the start-up and also be used for organizational improvement.

As a result of the start-up of the new unit and the survey-feedback interventions that took place, a job-redesign program was eventually undertaken in Unit II.

Thus, the research team was given an opportunity to compare the impacts of survey-feedback, job-redesign, and sociotechnical system interventions on employee satisfaction and productivity. The design of the study is shown in Figure 1.

During the first time period, the research team administered surveys in both units. At that time, employees who were to staff the redesigned unit were still in their old jobs at the site, working under the traditional system of management.

	Time 1 1974	Time 2 1975			Time 3 1976		
UNIT I	0_1	X_1	X_2	0_2	X_3		0_3
UNIT II	0_1	X_1		0_2	X_3	X_4	0_3

KEY:

$0_{1\,2\,3}$ = Survey-Research Measures
X_1 = Survey Feedback (waterfall method)
X_2 = Sociotechnical System Redesign of Unit I to Process Product T
X_3 = Survey Feedback (bottom-up method)
X_4 = Job Redesign in Unit II

Figure 1. Experimental Design

The second period began with the feedback of the data to site personnel. Beginning with the site manager, each supervisor met with his or her subordinates to share the data and solicit suggestions for improvement. The researchers were present throughout the feedback process to answer questions about the data and to facilitate discussion.

The sociotechnical system redesign of Unit I also took place during the second time period. The design of the unit incorporated many of the key features mentioned by Walton (1972), including autonomous (self-directing) work groups, integrated support functions, challenging job assignments, job mobility, facilitative leadership, managerial decision information for operators, and opportunities for continuous learning and evolution. The design of the production process itself called for several production stages, each separated by technologies for storing buffer inventories. Each stage of the process was to be operated by an autonomous work group in which all group members would eventually become skilled at all of the tasks in their areas. In addition, these groups would perform periodic maintenance, quality control, training, and employee-selection functions. To encourage team effort, all employees received a single pay rate. The second period ended with the administration of another survey measure.

The third period began with the feedback of the data from the second survey, this time beginning at the lowest level of the organization. Each supervisor and his or her subordinates discussed the data and formulated action proposals to be considered by the next level of management. Through this process, top management eventually received a list of employee-initiated suggestions for consideration and action.

A short time later, as a result of the redesign of Unit I and the survey-feedback process, a job-redesign program was undertaken in Unit II that resulted in greater variety and responsibility for a number of employees. A rotational process was begun in some areas, and some of the most repetitive jobs were eliminated through automation or divided among a number of employees.

Unlike in Unit I, there were no efforts in Unit II to form autonomous work groups, alter the existing technology, or change the style of management. Individual operators were given responsibility for the performance of some maintenance and quality-control functions, and a small pay raise was granted to equalize conditions across the site. The study was concluded with the administration of a third survey measure in both units.

A comparison of time 1 versus time 2 data in Figure 1 indicates the effects of the sociotechnical system and survey-feedback interventions that occurred in Unit I versus the survey-feedback intervention alone that occurred in Unit II. Similarly, a comparison of time 2 versus time 3 data reveals the effects of the second survey-feedback intervention in Unit I versus the survey-feedback and job-redesign interventions that occurred in Unit II. Finally, a comparison of time 1 versus time 3 data contrasts the effects of the overall combined interventions in both units.

Results of the Study

Both attitudinal and productivity measures were used to evaluate the impacts of the interventions. At each survey administration, a full sampling of the hourly population was sought, although completion of the surveys was voluntary. Company time was provided for survey completion, and employees were assured that their responses would be completely anonymous. In all, 195 employees and supervisors completed the first survey, 209 the second, and 184 the third. Although the number of persons at the site fluctuated over time, at least 90 percent of the available hourly population completed each survey.

Because the research was concerned with employee responses to the interventions, supervisory data were excluded from the analysis. Also, some surveys were not usable due to substantial missing information. This left 175 first surveys, 128 second surveys, and 154 third surveys for analysis.

A complete description of the survey measures is provided in Pasmore (1976). Ten measures of employee attitudes were included in the surveys: general job satisfaction, alienation (Blauner, 1964); job involvement (Lodahl & Kejner, 1965); intrinsic motivation (Hackman & Lawler, 1971); need strength (Hackman & Lawler, 1971); task attributes of employee jobs (Turner & Lawrence, 1965; Hackman & Oldham, 1974); measures of satisfaction with specific aspects of working conditions; supervisory consideration (Fleishman, 1960); organization structure (Friedlander, 1973); and intergroup relations. The average internal consistency reliability coefficients for these measures across the three survey administrations ranged from .53 to .89 with an average of .70.[1]

In addition to the survey measures, records of productivity, start-up costs, absenteeism, and on-site observations were kept during the study.

The effect of the intervention on the means of the survey variables in each unit over time is presented in Table 1. Univariate and multivariate analyses of variance performed on the data indicated that overall inter-

[1] All internal consistency reliability coefficients were computed using Cronbach's Alpha (1951).

Table 1. Means and Standard Deviations of Survey Variables by Unit over Time

Variable	Time 1		Time 2		Time 3	
	Unit I N = 61	Unit II N = 114	Unit I N = 42	Unit II N = 86	Unit I N = 56	Unit II N = 98
General Job Satisfaction	4.164 * (1.734)**	4.272 (1.821)	5.643 (1.495)	4.744 (2.019)	5.482 (1.250)	4.980 (1.250)
Specific Job Satisfactions	3.736 (1.195)	3.735 (1.186)	4.301 (.860)	3.487 (1.144)	3.937 (.854)	3.947 (.911)
Alienation	3.235 (1.118)	3.298 (1.105)	1.500 (1.358)	1.942 (1.577)	1.589 (1.255)	1.758 (1.429)
Job Involvement	3.650 (1.517)	3.544 (1.307)	4.738 (1.726)	4.767 (2.027)	4.875 (1.926)	4.765 (1.827)
Intrinsic Motivation	5.585 (1.186)	5.526 (1.225)	6.000 (1.249)	5.733 (1.718)	5.964 (1.464)	5.908 (1.437)
Intergroup Relations	3.574 (1.283)	3.789 (1.065)	3.764 (1.213)	3.708 (1.256)	3.482 (1.073)	3.506 (1.051)
Organic Structure	2.658 (1.239)	2.759 (1.103)	4.678 (1.034)	4.139 (1.040)	4.357 (.794)	4.290 (.994)
Supervisory Consideration	3.951 (1.374)	4.250 (1.268)	4.282 (1.397)	3.446 (1.540)	3.708 (.912)	3.571 (1.184)
Job Design	4.121 (.936)	4.116 (.824)	4.963 (1.020)	4.733 (1.086)	4.701 (.903)	4.735 (.936)
Absenteeism	2.033 (1.402)	3.509 (2.465)	2.095 (4.762)	2.663 (5.233)	2.750 (2.158)	3.102 (3.892)
Need Strength	5.708 (.855)	5.700 (.849)	N.A.	N.A.	5.116 (.975)	5.463 (.943)

*The first figure reported is the mean.
**The second figure (in parentheses) is the standard deviation.

Key: 1 = Low
7 = High

ventions in both units affected a number of employee attitudes positively and significantly (Pasmore, 1976). Significant ($p < .01$) increases were noted in the levels of general job satisfaction, intrinsic motivation, and job involvement; alienation decreased markedly ($p < .01$). Job design was reported to be more complex ($p < .01$), and the organization was perceived to have acquired a more flexible structure ($p < .01$).

Thus hypothesis 1 was supported, as both human-process and technostructural interventions did result in improvements in employee attitudes. It is interesting to note the different trends in employee attitudes in each unit. In Unit I, the initial combined sociotechnical system and survey-feedback intervention resulted in greatly improved employee attitudes. The survey-feedback intervention that occurred later in Unit I did little to further improve employee attitudes. In Unit II, on the other hand, the initial survey-feedback intervention alone resulted in improved employee attitudes, and the later combined survey-feedback and job-redesign interventions resulted in further improvements. Thus, it seems that *combined* human-process and technostructural interventions have a somewhat greater impact on employee attitudes than human-process interventions alone.

In Unit I, the effects of the sociotechnical system and survey-feedback interventions on the planned versus actual productivity of the operation were quite dramatic. Based upon start-ups of similar facilities elsewhere in the corporation, the original cumulative start-up production volume for Unit I was set at 1,093,600 units. By the end of the start-up period, approximately 1,450,000 units had been produced, or 133 percent of that originally planned. Despite the increased volume and corresponding raw material costs, the start-up cost was less than predicted. Instead of an estimated eleven-month operating cost of $931,800, the cost was approximately $860,000, a savings of $71,800, or 7.7 percent of the planned amount. The major saving achieved during the start-up was in terms of labor cost due to the sociotechnical system design of the unit. Industrial-engineered standards called for a crew of between 126 and 129 employees to operate the facility, while the actual number required under the new design was 104. A conservative estimate places the annual fixed labor expense savings at $264,000. It should be noted that these results were achieved with a unionized labor force and by adapting the optimal design of the unit to fit the existing facilities at the site.

The impact of the survey-feedback and job-redesign interventions in Unit II was not as impressive. Although the productivity of the two units could not be compared directly because they were processing different products, data obtained from the company records indicated no significant changes in productivity that could be attributed to the interventions in Unit II. It should be noted that the technology of the Unit II operation was complex and constrained the rate of productivity. Perhaps because the technology in Unit II was not affected greatly by the job-redesign or survey-feedback interventions, no improvement in productivity could have taken place. Other operating data collected by the Unit II supervisors indicated an improvement from time 2 to time 3 in terms of equipment downtime ($t = 4.61$; $p < .01$; $df = 33$; two-tailed test) and an increase in the time spent in meetings with employees ($t = 2.57$; $p < .05$; $df = 30$). There was no significant change in the number of cases of product that had to be reworked, a measure of quality. The job-redesign intervention did reduce the number of employees needed to operate the facility (by approximately six out of eighty). Thus, Unit II was able to maintain its production levels with fewer employees, despite the fact that more time was spent in meetings.

In summary, although the interventions had similar effects in terms of improving employee attitudes, only the sociotechnical system intervention was associated with significant improvements in productivity and cost savings.

Discussion

In interpreting these findings, it would be easy to jump to the conclusion that sociotechnical system interventions should be applied in all organizations and other techniques should be abandoned. This certainly is not the case. Even in this capital-intensive setting, the value of the human-process/survey-feedback intervention was apparent to the research team. Not only did it offer the means to collect important information regarding employee sentiments at the site, but it also became a vehicle for building trust between labor and management. Prior to the study, relationships between union and management at the site were typically adversarial, and there was some question in the minds of both the managers and researchers about the role the union would play in the study. It was not surprising to find that the union resisted the study and the improvement program at the site initially. Union members claimed that the improvement program was another effort by management to "get more work out of employees for the same pay" and that the research study was being conducted to back up the actions taken by management. These rumors were dispelled only after long hours of discussion in survey-feedback meetings and subsequent demonstrations by management of their good faith in responding to employee concerns.

Although the sociotechnical system method is intended to provide better working conditions for employees as well as greater productivity, in practice it easily could be perceived by employees as another management gimmick.

To dispel this notion, it is essential that employees have an active part in shaping their work environment, and it is here that human-process interventions are most useful. In retrospect, it would seem that combined human-process and technostructural interventions present the most balanced and potentially effective approach to organizational change in this type of setting. Managers and experts need to overcome the temptation to apply a neat and quick technique. If the results of this study can be generalized, as we believe they can, organizational improvement will be the result of a concerted effort to change the actual jobs of employees, with their participation, through changes in the technology. One-sided efforts aimed either at smoothing over employee attitudes or simply altering technological arrangements will probably meet with resistance and do more harm than good.

In other settings that do not involve capital-intensive technologies, such as hospitals, universities, or government agencies, human-process interventions may have more impact than technostructural techniques. To date, sociotechnical system and job-redesign theorists have not offered much guidance for change in this type of setting. We would expect that even limited technostructural interventions would produce more positive effects when combined with human-process methods in such settings than would human-process interventions alone. Simply changing employee attitudes is probably *not* enough to improve productivity. More comparative studies are necessary to determine the most effective combinations of interventions.

REFERENCES

Blauner, R. *Alienation and freedom*. Chicago: University of Chicago Press, 1964.

Cronbach, L. Coefficient Alpha and the internal structure. *Psychometrika*, 1951, *16*, 297-334.

Fleishman, E. *Manual for the leadership opinion questionnaire*. Chicago: Science Research Associates, 1960.

Friedlander, F. *Organization structure inventory*. Cleveland: School of Management, Case Western Reserve University, 1973.

Friedlander, F., & Brown, L. D. Organization development. *Annual Review of Psychology*, 1974, *25*, 313-341.

Hackman, J., & Lawler, E. Employee reactions to job characteristics. *Journal of Applied Psychology*, 1971, *55*, 259-286.

Hackman, J., & Oldham, G. *The job diagnostic survey: An instrument for the diagnosis of jobs and evaluation of job redesign projects*. New Haven, Conn.: Technical Report #4, Department of Administrative Sciences, Yale University, 1974.

Kahn, R. L. Organization development: Some problems and proposals. *Journal of Applied Behavioral Science*, 1974, *10*, 485-502.

Lodahl, T., & Kejner, M. The definition and measurement of job involvement. *Journal of Applied Psychology*, 1965, *49*, 24-33.

Pasmore, W. *Understanding organizational change: A longitudinal investigation of the effects of sociotechnical system, job redesign, and survey feedback interventions on organizational task accomplishment and human fulfillment*. Unpublished doctoral thesis, Purdue University, 1976.

Srivastva, S., Salipante, P., Cummings, T., Notz, W., Bigelow, J., & Waters, J. *Job satisfaction and productivity*. Cleveland: Case Western Reserve University, 1975.

Turner, A., & Lawrence, P. *Industrial jobs and the worker: An investigation of response to task attributes*. Boston: Harvard University, Graduate School of Business, 1965.

Walton, R. E. How to counter alienation in the plant. *Harvard Business Review*, 1972, *50*, 70-81.

Sociotechnical Systems: Some Suggestions for Future Research

Michael K. McCuddy

Since Trist and Bamforth (1951) published their seminal study of British coal mining operations, many scholars and researchers of organizational theory and behavior have turned their attentions to sociotechnical systems. A sampling of the studies that were fostered by Trist and Bamforth's data demonstrates the diversity that has occurred in sociotechnical system research. Some studies have addressed relationships between technological routineness and measures of organization structure or social structure (Mohr, 1971; Hage & Aiken, 1969; Perrow, 1967) or between technological complexity and span of control (Bell, 1967). Other research has been directed at classifying different types or varieties of technical systems (Thompson, 1967; Woodward, 1965). Still others have investigated the impact of technological change on the organization's social system (Burack & Cassell, 1967; Rice, 1958). Scholars have also addressed the impact of technology on the decision-making autonomy of lower level management personnel (Grimes & Klein, 1973). One study investigated the probable relationships among human operators, work tasks, and technology in future industrial production systems (Cooper, 1972).

The diversity in sociotechnical system research has generated a need to take stock of where sociotechnical research has been and the direction it should take in the future. This paper reviews the state of current sociotechnical knowledge and suggests some appropriate avenues for future research.

Five related concerns are addressed:

1. What is technology and how is it measured?

2. How does one delineate the social system in organizations, and what are the important characteristics of organizational social systems?
3. What are the important relationships between social and technical systems in organizations, and what limitations does the technical system impose on the social system?
4. To what extent and in what way is organizational effectiveness influenced by organizations' sociotechnical systems?
5. What improvements are needed in the current methods of conducting research on sociotechnical systems?

TECHNOLOGY CONSTRUCT

One crucial issue for students of sociotechnical systems is confusion about the meaning and content of the technology construct. This confusion stems from different concepts or ideas of what technology is, i.e., differing theoretical and operational definitions of technology, and from the multidimensionality of the construct.

Definitions of Technology

Technology can be defined in narrow or in broad terms. The narrowest definition encompasses *machine* technology, "the mechanical means for replacing human effort and for producing goods and for producing goods and services" (Kast & Rosenzweig, 1974, pp. 181-182). Broad general definitions of technology are based on the idea that technology represents knowledge about how work is accomplished. For example, Mesthene (1970, p. 25) defined technology as "the organization of knowledge for the achievement of practical purposes," including machines and intellectual tools such as computer languages and contemporary mathematical and analytical techniques. Dubin (1968, p. 467) also defined technology in a broad sense as the tools, instruments, machines, and technical formulas necessary to accomplish the work, as well as the body of knowledge that expresses the objectives of the work, its functional importance, and the rationale for the methods used. Although Dubin's definition is more specific than Mesthene's, both are nebulous and lack precision. Yet, both represent well-accepted conceptualizations of technology.

Technology also has been defined in terms of process. Hunt (1970, p. 239) saw technology as a transformation process in which "various things are done, with or without tools and machines, to transform inputs into outputs"—again, a very nebulous, imprecise definition.

Burack (1966, p. 47) had a similar view, but he focused on industrial production and defined process technology as "a systematic series of actions transforming materials into semifinished or finished products based on scientific concepts as applied to the industrial arts." He gave five underlying components of process technology: degree of mechanization, degree of time interdependence, degree of control instrumentation, degree of subdivision of labor, and degree of technical engineering organization of the process.

Other scholars have attempted to define technology by categorizing different types or varieties. Hickson, Pugh, and Pheysey (1969), for example, proposed a classification scheme of operations technology, materials technology, and knowledge technology. Woodward (1965) used technical complexity as a means of classifying unit and small-batch production, large-batch and mass production, and process production. Thompson (1967) categorized technologies as long-linked, mediating, or intensive.

These diverse conceptualizations present at least two dilemmas for the student. First, it is unclear what is meant by technology. Present definitions tend to be vague and imprecise and represent widely varied ideas. Second, advances in the comparative analysis of organizations' sociotechnical systems may be impeded by the lack of a precise and consistent conceptualization that is applicable to a wide variety of organizations. As Mohr (1971, p. 446) pointed out, "technology is . . . only a very broad concept that must become more specific to be useful for research and theory."

Multidimensionality of the Concept

Several dimensions of the technology construct have been written up in the literature. These include: technical or job complexity (Zwerman, 1970; Bell, 1967; Woodward, 1965); routineness of work (Hage & Aiken, 1969; Perrow, 1967); uniformity of tasks (Litwak, 1961); variety in the production system (Rackham & Woodward, 1970); and certainty of the task environment (Morse, 1970; Lawrence & Lorsch, 1967). Unfortunately, these measures are unidimensional, when in reality technology is a multidimensional construct.

Hrebiniak (1974), Lynch (1974), and Mohr (1971), among others, reported difficulty defining the construct of technology because of its multidimensionality. Yet, before some recent studies, little effort was made to ascertain the several important dimensions of technology. Lynch (1974) identified three independent dimensions of technology in a second-order factor analysis of data collected from functional departments of academic libraries; these factors were overall routineness, interdepartmental task interdependence, and library technology. Overton, Schneck, and Hazlett (1977) extracted three orthogonal factors—uncertainty, instability, and variability—from a thirty-four-item questionnaire that measured technology. Additional research is needed to describe technology completely.

THE CONCEPT OF A SOCIAL SYSTEM

Sociotechnical system researchers also need to define the characteristics and boundaries of organizational social systems. Several important questions need to be answered: What components, variables, and processes are included in and excluded from organizations' social systems? Which of these are most salient for different types of technical systems? Which of these affect or are affected by technology, and which are independent of technology (cf. Rice, 1958, p. 4)? To what extent should psychological properties be considered when defining an organization's social system?

Attempts to clarify characteristics of social systems within organizations are often imprecise. Litterer (1973, p. 284), for instance, referred to the social system as "the web of behavior of the individuals that operate the work or technical system." But the question really is what does this web of behavior include? Both the psychological and sociological aspects of behavior? Style of supervision? Communications and interaction-influence patterns of the management system? It has never been made clear what is and what is not included in an organization's social system.

Albanese (1975), focusing on the social aspects of designing sociotechnical systems, suggested that consideration be given to (1) the specific behaviors that are necessary to achieve system goals, (2) the consequences to the system members of executing or not executing the desired behaviors and actions, and (3) the means by which consequences are related to desired behaviors (p. 117). The first two imply that eliciting desired behaviors or planning and controlling human behavior (both traditional management functions) are components of organizational social systems. Scholars must ascertain the extent to which this is true. The second and third considerations are indicative of reward and punishment systems within organizations. The appropriateness of including these in a conceptualization of organizational social systems also must be established.

A different component of some conceptualizations of social systems is "role." Katz and Kahn (1966, pp. 37-38) cited three major social system components: "the *role* behaviors of members, the *norms* prescribing and sanctioning these behaviors and the *values* in which the norms are embedded." The role concept was also central to Hill's (1971, pp. 233-237) idea of organizational social systems. Hill also suggested that traditional structural concepts such as number of levels in the organizational hierarchy and span of control be included. Hage and Aiken (1969) also promoted centralization and formalization as elements of social structure. Jelinek (1977) also considered the social system to be a structural system. Some questions for further research include: Whether traditional structural concepts should be included in the social system framework, and whether organization structure and organizational social systems are equivalent concepts; if not, whether there are any similarities between structure and the social system and what these similarities are.

In summary, existing knowledge about organizational social systems needs to be extended in several ways. First, boundaries of the social system must be established by ascertaining which variables should be included. Second, as Hage and Aiken (1969, p. 366) have pointed out, an accurate means for measuring these components needs to be designed. Finally, research must be directed at determining which characteristics of social systems operate independently of technology and which characteristics interact with technical systems.

MERGING SOCIAL AND TECHNICAL CONSIDERATIONS

Although considering social and technical systems separately is a significant area for future research, considering their interrelationships is equally promising.

There is general agreement that an organization's technical system limits and restricts the accompanying social system (Litterer, 1973; Katz & Kahn, 1966; Rice, 1958), but there is disagreement about the extent of that limitation. According to Katz and Kahn (1966, p. 433), some technical systems require a specific type of social system and other technical systems are compatible with several *different* social arrangements. However, Litterer (1973, p. 284) reported that "there are several forms of social systems that would be compatible [*with*] and supportive of a particular technical system: technology sets the limits for a range of possible social forms." These conflicting findings suggest the need to ascertain how and to what extent organizations' technical systems limit or constrain their social systems and, conversely, whether particular types of social systems place limitations on the type of technical system used in an organization. Research should also focus on the relative compatibility of different types of technical and social systems.

Some authors have noted that technological variables and social structure variables are frequently confused (Stanfield, 1976; Gillespie &

Mileti, 1977; Jelinek, 1977). They lament the inadequate designation of specific variables that represent each system. This suggests that future research should be done to work out separate constructs for each and to avoid overlap.

Another current area for research is the study of moderating variables in sociotechnical relationships. Reeves and Woodward (1970) proposed that the nature of the control system may moderate the relationship between technology and organizational behavior. Jelinek (1977) proposed "intervening technologies" and structure as moderators of the relationships between "core technology" and the organizational environment. Closeness to technological change was used as a moderator variable in research reported by Billings, Klimoski, and Breaugh (1977). In general, however, little is known and much must be learned about the effects of moderating variables in sociotechnical systems.

EFFECTIVENESS AND SOCIOTECHNICAL CONSONANCE

Research by Burns and Stalker (1961), Woodward (1965), Perrow (1967), and Nemiroff (1975) suggested that organizations will perform effectively only to the extent that their structures are compatible with the requirements and dictates of their technical systems. This is generally known as the *consonance hypothesis* (Mohr, 1971). Burns and Stalker (1961) found that mechanistic systems of management were suitable for firms with stable technologies and environments and that organic managerial systems were appropriate for organizations with rapidly changing environments and technologies. They also reported the difficulties experienced by organizations that attempted to impose organic management systems on stable and unchanging technologies and environments or attempted to retain mechanistic managerial systems when their technologies and environments were changing rapidly. Perrow (1967) suggested that organizations probably would operate most effectively with a bureaucratic structure when doing routine tasks. Woodward (1965) discovered that those firms she labeled "above average in success" tended to have different structural characteristics from those firms she designated "below average in success." Nemiroff (1975) found that high task effectiveness was associated with consonance between a work unit's task and its organizational structure.

Many industrial studies and organizational change efforts also tend to support this hypothesis. Trist and Bamforth's (1951) data showed that productivity and worker satisfaction declined and absenteeism increased when British coal mines changed to the conventional longwall method of mining. They explained this as the result of a lack of consonance between the longwall technology and the pre-existing social structure of the miners. Rice's (1958) studies of Indian textile mills also demonstrated the individual and organizational benefits of having compatible social and technical systems. Industrial experiments at Shell UK Limited (Hill, 1971), the Volvo Kalmar Plant, and a pet food plant (Walton, 1972) also suggested the benefits of considering both social and technical aspects when designing effective organizational systems.

Mohr (1971), however, was not supportive of the consonance hypothesis. In a recent study of technology and social structure (1971), he examined effectiveness in terms of the consonance between two technology variables (routineness and task interdependence) and one social structure dimension (participativeness of supervisory style). He found no support for the proposition that organizational effectiveness is determined by the compatibility of social structure and technology.

These results do raise some significant issues for sociotechnical researchers. In particular, the validity of the consonance hypothesis needs to be explored thoroughly. Organizational effectiveness may be related to some social and technical variables but not to others. Highly mechanistic or highly organic structures may "require" a consonant social structure; structures closer to the mean or "mixed" structures may not require consonance to be effective. In the case of "mixed" structures (those which are neither highly mechanistic nor highly organic), one social system may be as appropriate as any other in terms of organizational effectiveness.

Research is needed to determine which social and technical variables must be consonant for organizations to be effective and whether these variables vary from situation to situation. Future studies are needed to examine the extent to which sociotechnical consonance determines organizational effectiveness and to find other organizational variables that may be related to effectiveness.

Researchers must also address the concept of organizational effectiveness itself. Beckhard (1969) pointed out the multidimensionality of the concept by listing ten different components in his definition. Mahoney (1967) listed twenty-four basic dimensions of organizational effectiveness. This multidimensionality affects knowledge about sociotechnical systems. Researchers must determine which components of organizational effectiveness are related to the sociotechnical system and investigate the feasibility of a contingency approach to the relationship. There is a need to establish whether different aspects of effectiveness are related to sociotechnical consonance depending on the different sociotechnical system being studied.

METHODOLOGICAL CONSIDERATIONS

Researchers should also examine their methodologies. A good deal of the existing research consists of case studies (Burack & Cassell, 1967; Rice, 1958; Trist & Bamforth, 1951) or correlational studies (Woodward, 1965; Mohr, 1971; Child & Mansfield, 1972; Khandwalla, 1974; Comstock & Scott, 1977). These do not determine causality or rule out threats to external validity (Campbell & Stanley, 1963). Appropriately designed and executed studies would help determine causality and eliminate alternate explanations for research results. Laboratory and field experiments and quasi-experimental designs (Campbell & Stanley, 1963) are potentially valuable tools for research (Cummings, 1976; Billings, Klimoski, & Breaugh, 1977).

Technology has been extensively researched at the individual and group levels, but only a few studies have focused on technology at the organizational level (Hickson, Pugh, & Pheysey, 1969). Consequently, there is also a need for research using the organization as the unit of analysis. Also, researchers must develop adequate operational measures of sociotechnical variables.

SUMMARY

This paper provides an overview of the current status of research and knowledge about sociotechnical systems and related organizational phenomena. A review of the literature indicates the need for additional research on: the *technology construct*, the development of a clear conceptualization of *social systems* in organizations, the *interrelationships* of social and technical systems, the relationship between *organizational effectiveness* and sociotechnical consonance, and the *methodology* of sociotechnical system *research*.

Several aspects of the technology construct are in need of investigation. The concept of technology must be refined, defined explicitly, and made more specific, particularly for comparative analysis of sociotechnical systems. Scholars should also ascertain which dimensions of technology are most useful for sociotechnical system research and theory and develop adequate measures for these dimensions.

It is imperative that we know what a social system within an organization encompasses. Adequate and accurate measures of the social system's characteristics must be developed, and students must determine which social system dimensions are important in their own right and which are important because of their interrelationships with technical dimensions.

Future research must also address the interrelationships of social and technical systems in organizations. The way(s) in which technical systems constrain or limit social systems and vice versa must be investigated. Knowledge also can be generated concerning (1) the types of social systems that are compatible with specific technical systems; (2) the importance of technology, relative to other organization factors, in determining and limiting the social system; and (3) the presence of moderating variables in sociotechnical relationships.

The relationship between organizational effectiveness and sociotechnical consonance represents a promising area for research. The validity of the consonance hypothesis and the extent to which sociotechnical consonance influences organizational effectiveness can be investigated. Students of sociotechnical systems also must examine the interrelationships among different characteristics or dimensions of the social system, the technical system, and organizational effectiveness in diverse organizational settings.

Sociotechnical research currently suffers from a lack of methodological variety, especially in terms of eliminating alternate explanations for research findings and ascertaining the sources of causality in sociotechnical systems. In the future, those who design sociotechnical research projects could use a wider array of experimental and field methodologies and focus on entire organizations as well as on individuals and groups.

REFERENCES

Albanese, R. *Management: Toward accountability for performance*. Homewood, Ill.: Richard D. Irwin, 1975.

Beckhard, R. *Organization development: Strategies and models*. Reading, Mass.: Addison-Wesley, 1969.

Bell, G.D. Determinants of span of control. *American Journal of Sociology*, 1967, *73*, 90-101.

Billings, R. S., Klimoski, R. J., & Breaugh, J. A. The impact of a change in technology on job characteristics: A quasi-experiment. *Administrative Science Quarterly*, 1977, *22*, 318-339.

Burack, E. H. Technology and some aspects of industrial supervision: A model building approach. *Academy of Management Journal*, 1966, *9*, 43-66.

Burack, E. H., & Cassell, F. H. Technological change and manpower developments in advanced production systems. *Academy of Management Journal*, 1967, *10*, 293-308.

Burns, T., & Stalker, G. M. *The management of innovation*. London: Tavistock, 1961.

Campbell, D. T., & Stanley, J. C. *Experimental and quasi-experimental designs for research*. Chicago: Rand McNally, 1963.

Child, J., & Mansfield, R. Technology, size and organization structure. *Sociology*, 1972, *6*, 369-393.

Comstock, D. E., & Scott, W. R. Technology and the structure of subunits: Distinguishing individual and workgroup effects. *Administrative Science Quarterly*, 1977, *22*, 177-202.

Cooper, R. Man, task and technology: Three variables in search of a future. *Human Relations*, 1972, *25*(2), 131-157.

Cummings, T . G. Socio-technical systems: An intervention strategy. In W. Warner Burke, *Current issues and strategies in organization development*. New York: Human Service Press, 1976.

Dubin, R. *Human relations in administration* (3rd ed.). Englewood Cliffs, N.J.: Prentice-Hall, 1968.

Gillespie, D. F., & Mileti, D. S. Technology and the study of organizations: An overview and appraisal. *Academy of Management Review*, 1977, *2*(1), 7-16.

Grimes, A. J., & Klein, S. M. The technological imperative: The relative impact of task unit, modal technology, and hierarchy on structure. *Academy of Management Journal*, 1973, *16*, 583-597.

Hage, J., & Aiken, M. Routine technology, social structure, and organization goals. *Administrative Science Quarterly*, 1969, *14*, 366-376.

Hickson, D. J., Pugh, D. S., & Pheysey, D. C. Operations technology and organization structure: An empirical reappraisal. *Administrative Science Quarterly*, 1969, *14*, 378-397.

Hill, P. *Towards a new philosophy of management*. New York: Barnes & Noble, 1971.

Hrebiniak, L. G. Job technology, supervision, and work-group structure. *Administrative Science Quarterly*, 1974, *19*, 395-410.

Hunt, R. G. Technology and organization. *Academy of Management Journal*, 1970, *13*, 235-252.

Jelinek, M. Technology, organizations, and contingency. *Academy of Management Review*, 1977, *2*(1), 17-26.

Kast, F. E., & Rosenzweig, J. E. *Organization and management: A systems approach* (2nd ed.). New York: McGraw-Hill, 1974.

Katz, D., & Kahn, R. L. *The social psychology of organizations*. New York: John Wiley, 1966.

Khandwalla, P. N. Mass output orientation of operations technology and organizational structure. *Administrative Science Quarterly*, 1974, *19*, 74-97.

Lawrence, P. R., & Lorsch, J. W. *Organization and environment*. Boston: Graduate School of Business Administration, Harvard University, 1967.

Litterer, J. A. *The analysis of organizations* (2nd ed.). New York: John Wiley, 1973.

Litwak, E. Models of bureaucracy which permit conflict. *American Journal of Sociology*, 1961, *67*, 177-184.

Lynch, B. P. An empirical assessment of Perrow's technology construct. *Administrative Science Quarterly*, 1974, *19*, 338-356.

Mahoney, T. A. Managerial perceptions of organizational effectiveness. *Management Science*, 1967, *14*(2), B-76–B-91.

Mesthene, E. G. *Technological change: Its impact on man and society*. Cambridge, Mass.: Harvard University Press, 1970.

Mohr, L. B. Organizational technology and organizational structure. *Administrative Science Quarterly*, 1971, *16*, 444-459.

Morse, J. A. Organizational characteristics and individual motivation. In J. W. Lorsch & P. R. Lawrence (Eds.), *Studies in organization design*. Homewood, Ill.: Richard D. Irwin, 1970.

Nemiroff, P. M. *The impact of individual, task and organization structure variables on task effectiveness and human fulfillment at a continuous automated-flow production site: A contingency approach*. Unpublished doctoral dissertation, Purdue University, 1975.

Overton, P., Schneck, R., & Hazlett, C. B. An empirical study of the technology of nursing subunits. *Administrative Science Quarterly*, 1977, *22*, 203-219.

Perrow, C. A framework for the comparative analysis of organizations. *American Sociological Review*, 1967, *32*, 194-208.

Rackham, J., & Woodward, J. The measurement of technical variables. In J. Woodward (Ed.), *Industrial organization: Behavior and control*. London: Oxford University Press, 1970.

Reeves, T. K., & Woodward, J. The study of managerial control. In J. Woodward (Ed.), *Industrial organization: Behavior and control*. London: Oxford University Press, 1970.

Rice, A. K. *Productivity and social organization: The Ahmedabad experiment*. London: Tavistock, 1958.

Stanfield, G. G. Technology and organization structure as theoretical categories. *Administrative Science Quarterly*, 1976, *21*, 489-493.

Thompson, J. D. *Organizations in action*. New York: McGraw-Hill, 1967.

Trist, E. L., & Bamforth, K. W. Some social and psychological consequences of the longwall method of coal-getting. *Human Relations*, 1951, *4*, 3-38.

Walton, R. E. How to counter alienation in the plant. *Harvard Business Review*, 1972, *50*(6), 70-81.

Woodward, J. *Industrial organization: Theory and practice*. London: Oxford University Press, 1965.

Zwerman, W. L. *New perspectives on organization theory*. Westport, Conn.: Greenwood Publishing, 1970.

PART VI:

PERSPECTIVES ON SOCIOTECHNICAL SYSTEMS

Introduction

The bold experiments conducted by General Foods, Volvo, Philips, and other companies are signs that the design of future work arrangements is already under way. The course of these experiments and others like them will do much to determine the expectations of employees about their jobs and of managers about the effective application of sociotechnical system techniques. It is important to consider what we have learned from experiments in the past. If sociotechnical system techniques are to be used to shape the nature of work in coming decades, we must take into account not only what these techniques have accomplished, but also what refinements are needed. Authors who have contributed their perspectives to this part of the book are Walton, Davis, Trist, Emery, and Pasmore, Srivastva, and Sherwood.

Walton reviews some current examples of sociotechnical system and union-management work-restructuring efforts. In his view, contemporary projects in the redesign of work have become more balanced, in that both productivity and quality-of-work-life improvements are being pursued simultaneously. He notes other trends in work-restructuring efforts: descriptions of both successes and failures are being reported more frequently; to assure success more time is being taken for restructuring; the theoretical assumptions underlying work restructuring are becoming more eclectic; there is greater diffusion of innovative structures across industries and work settings; and legislation increasingly embraces new standards of quality of work life. According to Walton, all of these factors are grounds for cautious optimism concerning the long-term growth and development of innovative work designs.

Davis discusses the relationships among sociotechnical system methods, the design of work, and the quality of working life. He notes a move away from the design of *individual jobs* to the consideration of larger *organizational and technical systems*. Inherent in this move is the realization that the technology used to manufacture goods and services is not predetermined, but is substantially influenced by the psychosocial assumptions of its designers, i.e., assumptions about human behavior, values, and the nature of organizational life. Davis concludes that as these trends intersect, sociotechnical system methods will provide new work structures that will bring improvements in the quality of working life.

Trist emphasizes the importance of collaboration in new work settings because of the need to deal with increasingly complex organizational environments. He states that collaboration will replace competition in the successful building of a post-industrial order that must deal with new levels of interdependence, complexity, and uncertainty. He calls for employee involvement in work-restructuring activities and describes collaborative efforts that have taken place at the work-group, organizational, and national levels. He notes that the pace of such efforts is increasing concomitant with the increase in environmental complexity.

Emery addresses that symbol of modern-day manufacturing—the assembly line. The logic underlying assembly-line design is discussed, and its inherent but unanticipated costs are noted. Emery points out that efforts to reduce labor costs and variability in the manufacturing processes require that designers cope with problems involving the transfer of materials, standardization of parts, balancing the work flow of operators, and supervision of the overall process. The aim of the designers has been least cost for the most product, and the assembly line was their way to achieve that goal. As Emery notes, however, fractionation of work leads to expenses associated with lowered human responsiveness, and extreme fractionation can lead to excessively high human costs. Although Emery does not take issue with the need to fractionate work, he does present the Volvo Kalmar experiment as an alternative to the assembly line as a way to meet the problems associated with fractionation. He concludes that the sociotechnical system design principles as they were used at Kalmar represent the best course for future work arrangements and recommends that traditional assembly-line technologies be abandoned.

Finally, Pasmore, Srivastva, and Sherwood point out that sociotechnical system applications, although successful, have been too narrowly confined to industrial settings. The authors propose a new, more encompassing theory intended to embrace work-restructuring attempts in both industrial and nonindustrial organizations: sociotask theory. The focus of sociotask theory is on the nature of tasks performed by people in organizations. The theory stresses the critical interdependence of tasks in organizations and the need for performance to be both dependable in outcome and equitable in reward. The authors contend that to affect the innovative restructuring of work organizations, the nature of tasks, technology, structure, and power relationships must be integrated effectively. Although the theory is in a developmental state, it does draw on new knowledge in the applied behavioral sciences that will be needed to shape future work arrangements.

In conclusion, the authors in this section are uniformly positive in their outlook regarding the continued application of sociotechnical system principles to the design of work. At the same time, they are aware that sociotechnical system theory, in its current state of development, is in need of improvement. We must come to grips with definitions of such terms as system boundaries, joint optimization, and environmental turbulence. We must develop better methods of measurement, diagnosis, and application if the use of these methods is to become less theoretical and more practical for utilization by management. Finally, we must learn more about the basic theories and assumptions, which will allow the diffusion of sociotechnical system methods to a variety of organizational settings. The tasks before us are not simple, but it is obvious that they must be addressed by researchers and by managers if the goal of designing work to satisfy both productivity and human needs is to be realized.

Perspectives on Work Restructuring

Richard E. Walton

The field of "work restructuring" has evolved during the last decade. In the late Sixties and early Seventies, scholars, managers, and journalists grew increasingly concerned about the "blue collar blues" and other more dramatic symptoms of alienation in the work place. From 1972 to 1975, scholars focused on the relatively few visible experimental solutions to worker disaffection available, e.g., the highly participative work system in General Food's plant at Topeka, Kansas; Volvo's pioneering modifications of the car assembly line; and the industrial democracy project at the Bolivar, Tennessee, plant of Harman Industries, sponsored jointly by management and the United Auto Workers. These were representative of perhaps another dozen similarly ambitious projects underway at that time, but they received a disproportionate amount of attention because observers with many different theories used the studies to illustrate their own particular ideals, aspirations, fears, and doubts.

At the present time, interest in and research about work restructuring have a broader base, are more balanced, are long term in conception, and are more eclectic in orientation. Each of these trends is discussed in the following paragraphs.

Work restructuring has a *broader base* in many respects. Probably the number of industrial firms in the United States with serious and significant work-restructuring programs increased more than tenfold from 1974 to 1977. A growing number of unions are participating with management in work-innovation projects. Several important nonprofit institutions, e.g., the Work in America Institute, the National Quality of

Work Center, and the Massachusetts Quality of Working Life Council, have been established to promote the work-reform movement. The field also has attracted an increasing amount of attention in the academic community.

Perhaps the most encouraging trend toward *balance* is related to the goals of work restructuring. Previously, many scholars and change agents conceived of their activities as part of a movement to improve quality of work life and regarded any direct concern about productivity as a distraction. At the same time, many managers assessed work-restructuring designs strictly in terms of their potential for improving productivity, without treating improvement in quality of work life as an important goal in its own right. Recently, both improvements in productivity (using the broadest conception of productivity) and improvement in quality of work life have been felt to be legitimate and urgent, and it has been agreed that these two goals not only can be, but should be, pursued simultaneously.

Balance in the literature is also being achieved; the trend toward reporting both successes and failures is a good example. For instance, a paper published in 1974 (Walton, 1974)[1] analyzed why so many projects languished or were terminated after an initial period of success. A later article (Walton, 1977)[2] analyzed why the Topeka plant discussed in an earlier study, although not without failures, was still robust after six years. Similarly, a study of diffusion efforts (Walton, 1975) focused on why these efforts failed so frequently; a 1977 study (Walton, 1977), however, analyzed three instructive examples of successful diffusion of work innovations through large multiplant firms.

Work-restructuring activities are increasingly occurring within a *long-term* framework. Previously, two views were prevalent: first, that work-restructuring innovation would spread rapidly from one plant to another, transforming the nature of the work place relatively quickly, and, second, that these innovations were merely a momentary fad with little general application. Both views are giving way to a third, often developed by top management after thoughtful examination of current activities and thinking in the field: that significant changes in the way work is organized and managed are not merely desirable, but inevitable, and that these changes will occur over a longer period of time.

Finally, the frameworks within which work restructuring takes place are increasingly *eclectic*. For example, there is less argument today about which is more important—extrinsic rewards (e.g., pay and

---

[1] See R. E. Walton, "Innovative Restructuring of Work" in Part VI of this text.

[2] See R. E. Walton, "How to Counter Alienation in the Plant" in Part IV of this text.

advancement) or intrinsic rewards (e.g., satisfaction generated by interesting work content). Increasingly, it has been recognized that both are important and that the emphasis each should receive in a restructuring effort depends on the particular work force, the nature of the tasks, and the economics of the plant. Managers, union officials, and change agents alike increasingly approach a given work situation without a preconceived theory or model of the form that change should take.

Although the above trends are positive ones that increase the soundness of the various approaches to restructuring activities, the author has several concerns about the field at this time. Too many projects launched under relatively favorable circumstances are bogging down during the implementation phase. One problem that comes up is a tendency toward either/or thinking, e.g., thinking that supervision must be either directive or nondirective (rather than seeing that supervision can be directive or nondirective depending on the circumstances). Another problem comes up if project leaders neglect to clarify forcefully the *quid pro quos* that underlie a participative work system, i.e., that freedom and influence need to be accompanied by responsibility; that an absence of external controls requires an assumption of greater self-discipline; and that a system that is extraordinarily attuned to the needs of its members will only be viable over time if its members are unusually attuned to the needs of the organization as an entity.

Although the outlook is optimistic for long-term growth and development of the innovative restructuring of work, we probably will continue to be confronted with a substantial number of failures as well as successes during the next few years.

A number of forces probably will promote the diffusion of the more effective innovative work structures from one industry to another, from blue-collar manufacturing work to white collar and service work, and from the private to the public sector. One of these forces is competition. Many firms in the United States have found that international competition requires them to pay attention to productivity. Another force is the changing expectations of workers. Workers' consciousness of quality-of-work-life issues will continue to rise. Another force is legislation that might set new, more embracing quality-of-work-environment standards or might encourage new forms of worker ownership or worker participation in the management of private firms. This legislation will be stronger in Canada than in the United States and will continue to be stronger in Europe than in North America. Nevertheless, there will be an increased level of U.S. governmental interest in work-restructuring activities. This is disconcerting because, of the three forces, it has the most potential for promoting ill-conceived and poorly implemented work-innovation projects.

REFERENCES

Walton, R. E. How to counter alienation in the plant. *Harvard Business Review*, Nov.-Dec., 1972, pp. 70-81

Walton, R. E. Innovative restructuring of work. In J. M. Rosow (Ed.,), *The worker on the job: Coping with change*. Englewood Cliffs, N.J.: Prentice-Hall, 1974.

Walton, R. E. The diffusion of new work structures: Explaining why success didn't take. *Organizational Dynamics*, Winter 1975, pp. 3-22.

Walton, R. E. Successful strategies for diffusing work innovations. *Journal of Contemporary Business*, June 1977, pp. 1-22.

Sociotechnical Systems: The Design of Work and Quality of Working Life

Louis E. Davis

In different but significant ways, sociotechnical systems are interrelated with the design of work and with the emerging concept of the "quality of working life." Although they started from different premises, the design of work, or job design, conceptually merged with sociotechnical systems through the collaborative work of Davis and Emery in the early Sixties. This union ultimately led to the evolution of the concept of quality of working life, with sociotechnical systems as one of its theoretical underpinnings.[1]

A sociotechnical system was conceived as an open system whose interacting social and technical systems were responsive to external influences, which permitted a re-examination of the crucial elements of work design. Consequently, emphasis shifted from concern about specific job design per se to concern about organization and technology as sociotechnical systems in which job designs were the dependent outcomes. This led to the views that jobs as organizational units were not conceptually appropriate bases for analysis, design, or redesign of work systems and that jobs were not appropriate units for making organizational changes that would enhance organizational effectiveness and the quality of working life. Nevertheless, most organizations in the United States continue to view jobs as a basis for organizational design and improvement. This approach is popular because in technologically advanced societies and organizations jobs are constantly changing, man-made inventions designed to suit a variety of

[1]For a review of the developments see Davis (1977).

technological and social systems needs. The sociotechnical systems framework gave rise to the idea that jobs are fundamentally derived from the design of larger organizational and technical systems; therefore, an examination of the larger systems (and their environments) reveals significant opportunities for change.

Also, the idea that jointly acting technical and social systems are the means to achieve an organization's objectives led to the realization that technological designers inevitably have certain assumptions and values about people and social systems. These values and assumptions are incorporated implicitly into the designs of technology, particularly production and computer data-processing constructs, which so centrally affect the quality of working life. Because this is true, those who design jobs must look at the historical design of production and computer data-processing technologies to see which social system planning and psychosocial assumptions were considered. They also must make explicit the economic and social factors included implicitly in the design process.

This implies two central realities about technology (Davis & Taylor, 1975).[2] First, the shape of technology is not predetermined by its own developmental "laws" (technological determinism) but is determined substantially by the psychosocial assumptions that its designers have incorporated. When these assumptions are seen as "given," the behavior of those interacting with the technology follows the only path available, thus amply fulfilling the prophecy implied in the design. Second, technological design incorporates assumptions about human behavior, relative value of men and machines, and the nature of the organization. Thus, in their organizational and technological designs, organizations reveal the values dominant in the society at the time the technology evolves. Once these realities are understood, the prospects for achieving an enhanced quality of life in the work place improve greatly, but a consideration of social system requirements necessary to achieve a high quality working life must be included in the design.

Two other consequences of research and design in sociotechnical systems are that designers work with larger organizational entities, with consequent greater complexity, and that they deal with the *roles* that employees play in the functioning of the organization, rather than with job descriptions. Since the late Sixties, the concept of role design (as distinct from the concept of job design) has become increasingly relevant. Learning how to design roles led progressively to the design of

---

[2]See L. E. Davis & J. C. Taylor, "Technology, Organization and Job Structure," reprinted in Part V of this text.

self-maintaining organizational units—the ultimate units or the building blocks of an organization.

Sociotechnically designed or redesigned organizations based on organizational units currently are being referred to as "new forms of work organization." When the design is coupled with the concept of "minimal critical specifications" (Cherns, 1976, p. 786)[3], i.e., specifying only those structural and role requirements that are crucial to coordination with and maintenance of the larger organization, then many of the goals of an enhanced quality of working life become possible to achieve.

In the design or redesign of organizations based on self-maintaining units, sociotechnical systems, work design, and quality of working life can be combined in the most powerful conceptual and practical ways to provide the means to satisfy individual needs and aspirations, social system needs, and organizational requirements.

Two issues that are important in the Seventies are (1) rapid change in the environments of organizations, which frequently requires great adaptability, and (2) continuing change in the values and expectations of those who work, which means that organizational design must provide for noncoercive, self-developed means to satisfy these changing needs. The concept of organization design is shifting toward an emphasis on cooperation and responsive adaptation; task integration; composite self-directing work groups; extended roles that provide high participation; wide competence; continued development; and appropriate work authority. This can be contrasted with the bureaucratic-scientific management concept of the directly controlled, one person-one task organizational unit. Sociotechnical approaches, combined with quality of working life goals, provide powerful comprehensive means for the "reform of work" or debureaucratization, which can lead to participative, developmental roles for members and noncoercive guiding, buffering roles for managers in evolving organizational settings.

REFERENCES

Cherns, A. B. The principles of sociotechnical design. *Human Relations*, 1976, *29*(8), 783-792.

Davis, L. E. Job design: Overview and future direction. *Journal of Contemporary Business*, June 1977, pp. 85-102.

---

[3]See A. B. Cherns, "The Principles of Socio-Technical Design" in Part III of this text.

Davis, L. E., & Taylor, J. C. Technology, organization, and job structure. In R. Dubin (Ed.), *Handbook of work, organization and society*. Chicago: Rand McNally, 1975.

Collaboration in Work Settings: A Personal Perspective

Eric L. Trist

Projections of what is likely to occur within the next 50 to 100 years, if substantial modifications are not made in advanced industrial societies, tend to be gloomy. It is as though once again a vision of a proximate end is beginning to reappear in human consciousness. At the same time a belief is also arising that any such dismal termination is far from inevitable and that active human intervention can prevent it.

I will attempt to show that institution-building (social architecture, to use Perlmutter's [1965] term) is of critical importance for this purpose; that the new organizational designs needed require to be based on collaborative principles; further, that the refashioning of work organizations in this idiom is the central institution-building task.

The prevailing organizational form in advanced industrial societies is the technocratic bureaucracy. This paradigm, as it has evolved historically, has been founded on the primacy of competition and win-lose relations. Internally, managers must compete for position and career opportunity, while the interests of the work force place it in an adversary

---

Eric Trist is Professor of Social Systems Sciences at the Wharton School, University of Pennsylvania. He was a founder-member of the Tavistock Institute of Human Relations in London and Chairman of its Human Resources Centre.

Reprinted from *The Journal of Applied Behavioral Science*, Volume 13, Number 3, 1977, pp. 268-278. Reproduced by special permission from NTL Institute.

position. Externally, the organization competes in the marketplace, if in the private sector; if in the public sector, the competition is with other departments—for budgets, territory, and personnel. The competitive technocratic bureaucracy is a singular organization supposed to have no other interest than its own self-interest. Collaboration is expected at the task level wherever tasks are interdependent, but this requirement stands in contradistinction to the psycho-political pressures on individuals and component groups. Conflict between these two aspects is endemic, despite efforts, however well intentioned and often partially successful, to reduce it.

One's approach to a theme such as collaboration in work settings will be very different according to whether one believes that the present industrial order based on the competitive and singular technocratic bureaucracy will continue indefinitely, with presumed beneficial consequences; or whether one believes that the risks of allowing it to do so are unacceptable—as a consequence of its increasing mismatch with emergent environmental processes and related signs of its own increasingly shaky performance. To those who hold the first position—and they are in the majority—the encouragement of collaboration in work settings is likely to appear peripheral, though possibly desirable as a means of raising productivity, or even on humanistic grounds. It will also tend to be regarded as a short-term tactical matter without historical meaning. To the minority who hold the second position it is more likely to appear as central and mandatory. To them it will tend to be seen as a critical factor in establishing the enabling conditions for human survival, to have long-term strategic implications, and to be an emergent social process of historical significance in the transition from an industrial to a postindustrial social order.

SOCIAL ARCHITECTURE AND CONTEXTUAL ENVIRONMENT

Some of the planners have now come to take this view. Ozbekhan (1971), in his concept of planning as a system of human action, has postulated organizational redesign on participatory lines as essential to the implementation process. Friedmann (1973) has suggested a rather simple concept of a "cellular" society that builds up "assemblies" in a Jeffersonian manner. Ackoff (1974) contends that organizations must pursue not only their own purposes (which have been their too exclusive concern in the past) but, equally, the purposes of their members—thence becoming "humanized." So also must they relate themselves to the purposes of the larger society—thence becoming "environmentalized." This model requires serious responsibility to be taken in the two critical domains neglected by the technocratic bureaucracy. It proposes a radical change in the historically evolved organizational stereotype.

Not many of those concerned with changes internal to the organization (changes at the micro level) have made the macro-social connection. Michael (1973) is one of the few writers who have bridged the gap between interpersonal relations and organizational development on the one hand and long-range social planning and what he calls "future-responsive societal learning" on the other.

Let me now state my own position. I include myself among those to whom unmodified persistence of the present industrial order is unacceptable, as I do not think this course is one which will work out. The redesign of conventional organizations seems to me to have become mandatory for any proactive attempt to bring into being successfully a future that will permit human survival under conditions worth having. I believe further that the enterprise of redesign involves more than marginal innovation. It requires a process of systemic transformation which will usher in a new paradigm. Though radical, both as regards values and forms, this transformation cannot be suddenly or easily attained. Arduous and protracted evolution, with much conflict and resistance, is rather to be expected. Moreover, the redesign enterprise will be successful only if it unites the micro and the macro perspectives. This involves on the one hand a process of organizational development that includes *work restructuring* and on the other hand a *planning process that is interactive and participatory*.

Since the present world has become interdependent on a scale hitherto unknown, this has the implication that collaboration, for the individual and the organization alike, has acquired primacy over competition. The many uses which the latter still retains have become subordinate.

Despite automation, the wholesale unemployment once expected has not appeared and the service sectors have grown. I therefore expect work organizations in their many different settings to remain central to society and to the individual. Change in this area has the potential for affecting numbers of people as no other kind of change has.

The reasons which have led me as a social scientist to hold this position derive from work begun with F. E. Emery as the decade of the 50s passed into the decade of the 60s. This work has continued in a number of studies, jointly or independently undertaken by us, all of which represent an attempt to relate the micro to the macro. In action research projects with which we were concerned, both our organizational clients and we ourselves were baffled by the extent to which the wider societal environment was moving in on their more immediate concerns, upsetting plans, preventing the achievement of operational goals, and causing additional stress and severe internal conflict. The extent of this environmental encroachment was recognized by those concerned as greater than that previously experienced. This phenomenon seemed to us to hold theoretical significance. Accordingly, we separated this wider environment, which we called the contextual, from the more immediate transactional environment and attempted a conceptual analysis of its characteristics.*

Four types of contextual environment were isolated. The first two, called *placid*, need not be discussed in the present context. They prevailed in pre-industrial societies where the change rate was slow. The third environmental type, however, called the *disturbed-reactive*, reflects an accelerating change rate and became increasingly salient as the Industrial Revolution progressed. This process zenithed some time after World War II when the science-based industries rose to prominence in the wake of the knowledge and information explosions. The disturbed-reactive character arose from the fact that the best chances of survival in this kind of world went to large-scale organizations with the capacity to make formidable competitive challenge through expertise and to maximize their independent power. The organizational form perfected for this purpose was that of the competitive and singular technocratic bureaucracy, in which the theories of Weber and Frederic Taylor are operationalized and matched to the requirements of the disturbed-reactive environment.

Organizations in a Turbulent Environment

The very success of the technocratic bureaucracy has given rise to a type of environment, very different from the disturbed-reactive, and with which this organizational form is mismatched. The new environment is called the turbulent field—in which large competing organizations, all acting independently in many diverse directions, produce unanticipated and dissonant consequences in the overall environment which they share. These mount as the common field becomes more densely occupied. The result is a kind of contextual commotion that makes it seem as if "the ground" were moving as well as the organizational actors. This is what is meant by *turbulence*. Subjectively it is experienced as "a loss of the stable state" (Schon, 1971).

---

*The main readings in this area are listed in the References at the end under Emery and under Trist.

As compared with the disturbed-reactive environment the turbulent field[1] is characterized by a higher level of *interdependence* among the "causal strands" (Chein, 1954) and a higher level of *complexity* as regards event-patterning. Together these generate a much higher level of *uncertainty*.

These higher levels of interdependence, complexity, and uncertainty pass the limits within which technocratic bureaucracies were designed to cope. Given its solely independent purposes, its primarily competitive relations, its mechanistic and authoritarian control structure, and its tendency to debase human resources, this organizational form cannot absorb environmental turbulence, far less reduce it. But such absorption and reduction are a necessary condition for opening the way to a viable human future, as the industrial gives place to the emerging postindustrial society.

THE PREMISES OF COLLABORATION IN NEW WORK SETTINGS

In Sartre's sense, the technocratic bureaucracy has been "depassed" in the historical process. Though Galbraith (1967) has referred to it—and the disturbed-reactive environment to which it is linked—as the "new industrial state," these are both better seen, McLuhan-wise, through the rear-view mirror, as comprising the old industrial state. Once one has become freed from past fixations in this regard, one is able to proceed with the evolution of values, cognitive orientations, and organizational modalities capable of matching up to the precarious state of affairs now looming in the contextual environment.

---

[1]The turbulent field has the characteristics of a richly joined environment in Ashby's (1960) sense. He did not think that the brain, as an ultra-stable system, could cope with such an environment. While this may be true in other species, the human brain, through its unusual capacity for abstraction from the concrete (Goldstein, 1939), is able to think in terms of "possible worlds." This enables man to be "ideal seeking," which Ackoff and Emery (1972) regard as the distinctively human attribute. The importance of ideals is that they can never be reached but provide continuous "guiding fictions" (Allport, 1937) in the pursuit of changing objectives and goals. Ideals are basic to value formation, and when common values are shared by large numbers of people they become able to undertake congruent courses of action. They can move in the same direction on the basis of "shared appreciations" (Vickers, 1965). These are independent of particular social structures. The adaptability imparted would appear to be basic for the capacity to cope with environmental turbulence. The most recent analysis of this is in Emery (1974).

Emergent social processes likely to have adaptive potential in this situation must be able to cope with the new levels of interdependence, complexity, and uncertainty. Collaboration rather than competition is a basic requirement for this purpose. It is as fundamental to the successful building of a postindustrial order as competition was to the successful building of an industrial order.

Acceptance of interdependence is founded on a willingness to align one's own purposes with those of diverse others and to negotiate mutually acceptable compromises rather than always trying to coerce and dominate in order to get one's own way. Collaboration in which "win-win" replaces "win-lose" situations, takes on the character of a primary value. Moreover, one has to respect and be concerned at the level of feeling with the others with whom one relates. This means recognizing qualities of the "heart" as well as the "head"—and the heart, as Maccoby (1976) has shown, is suppressed in the technocratic bureaucracy.

Complexity requires a conceptual strategy for problem solving different from and complementary to the prevailing analytic procedure which seeks to decompose wholes into parts. This is based on a synthetic procedure which involves relating parts to wholes. This process is intuitive as well as rational, and intuition is scarcely popular in the technocratic bureaucracy. Also, the order of complexity is apt to be too great for any one mind, so that emerging "appreciations" (Vickers, 1965) have to be built up in open interpersonal and group encounters. Once again collaboration is required.

Uncertainty puts a premium on innovation and the flexibility and free cognitive search required for it. The rigid structures and narrow range of approved and acceptable ideas in technocratic bureaucracies discourage creativity. As the processes involved are social, collaboration once again becomes basic but is inhibited by status barriers and the need always to secure personal credit.

It would seem that a reversal of conventional organizational attributes is required on all counts. Radical and systemic transformation becomes unavoidable, as the "organization character" (Selznick, 1957) itself needs to be changed.

Design Principles for a New Organizational Model

The embodiment of these new attributes in organizational forms poses a special problem which Emery (1967) approaches by postulating two design principles between which there is a choice. Whether for purposes of maintenance and renewal or of communication and adaptation, any organization requires some reserve capacity. In systems theoretic language this is called redundancy. Both design principles display redundancy, but in the first the redundancy is of parts and is mechanistic. The parts are broken down so that the ultimate elements are as simple and inexpensive as possible, as with the unskilled worker in a narrow job who is cheap to replace and who takes little time to train. The technocratic bureaucracy is founded on this type of design.

In the second design principle the redundancy is one of functions and is organic. Any component system has a repertoire which can be put to many uses, so that increased adaptive flexibility is acquired. While this is true at the biological level, as for example in the human body, it becomes far greater at the organizational level where the components—individual humans and groups of humans—are themselves purposeful systems. Humans have the capacity for self-regulation, so that control may become internal rather than external. Only organizations based on the redundancy of functions have the flexibility and innovative potential to give the possibility of adaptation to turbulent conditions. For it is through this redundancy of functions that their response repertoire acquires "the requisite variety"—in Ashby's (1960) sense—to meet the increased variation emanating from the environment.

From this analysis, a model would seem to be available, at least in outline, in terms of which organizations may be fashioned with some hope of surviving in a turbulent environment. They will tend to be socio-ecological rather than bureaucratic in their modes of regulation (Trist, 1976), with much local autonomy and a good deal of participation and democracy. Their parts will be mutually articulated rather than arranged in strict hierarchies. This change represents a move from a coercive towards a negotiated order, the establishment and maintenance of which depend on collaboration. Higher echelons will pursue missions in wider systems that lower echelons cannot undertake rather than preoccupy themselves with their internal workings. This represents a change in the concept of control from supervision to boundary maintenance. By simultaneously "environmentalizing" and "humanizing" organizations modelled in this way will become able to pursue their own purposes more effectively. Yet both internally and externally, relations will be founded on collaboration. Competition will be reserved for the marketplace and for such matters as the need to obtain alternative design and policy proposals and to have available more than one source of expert advice—which affects the public as well as the private sector.

A SURVEY OF EVOLVING COLLABORATIVE PROCESSES IN ORGANIZATIONS

Although no currently existing organization may be deemed to possess in a fully developed state all the attributes of the new model, many of the changes which have taken place in work organizations since World War II indicate that a number of partial moves in this direction have been successfully made. These have become more substantial in recent years.

A brief review proceeding from micro to macro may begin by noticing the extent to which toward the end of the 50s "organizational development" was taken up by key firms in the science-based industries in the United States. Belonging to the "leading part," such firms were the first to experience the pressures of turbulence (Trist, 1973). The accelerating change rate demanded flexibility and innovation. It was well therefore to encourage individuals to realize their potential. Also the complexity of the issues such firms face put a premium on team work and on more open interpersonal relations. The changes made were of climate rather than structure. The central question of power was not addressed (Bennis, 1970). The work force was untouched. Nevertheless, whatever the limits of early OD, the opening made was in the collaborative direction.

Job Satisfaction and Autonomous Workgroups

In the U.S. also, job breakdown was carried to such extremes after World War II that a counter-productive stage was reached (Walker & Guest, 1953), and the problem of alienation began to surface. Job enlargement and job rotation were tried and later job enrichment (Herzberg *et al.*, 1959), but it was restricted to individual jobs and management controlled. Participation was not allowed. This beginning, however, is to be regarded as significant in that it brought out the importance of job satisfaction as a critical factor in the humanization of the technocratic bureaucracy.

In Britain, also early in the postwar period, a new direction of development toward the new collaborative model began through the discovery of the autonomous workgroup.[2] This phenomenon gave rise to a new concept, the "sociotechnical system" (Trist & Bamforth, 1951), which represented a basic critique of scientific management in the

[2]No workgroup can, of course, be completely autonomous; semi-autonomous is frequently used; but Thompson's (1967) term, "conditionally autonomous," is more accurate.

technocratic bureaucracy. Sociotechnical theory is concerned with identifying the conditions which secure the best match between the social and technical systems. In the mining industry autonomous work-groups were traditional in surviving forms of unmechanized working. But they proved capable also of adapting themselves to semi- and to fully mechanized technologies. This was a most significant development for it represented an "organizational choice" (Trist *et al.*, 1963) in favor of the second design principle—the redundancy of functions. Autonomous groups have since emerged in continuous process industries, to which they are particularly suited, and have been experimented with in several other technologies. Despite their importance, they remain in themselves no more than component organizational processes, and thus represent micro changes.

Modeling Sociotechnical Principles in New and Established Plants

An operating plant is the smallest industrial system that can be regarded as an independent whole, even though it belongs to a larger corporate entity. A number of successful attempts to introduce sociotechnical principles to new plants have been made from the mid 60s onward. Early examples were a fertilizer plant in Norway (Emery & Thorsrud, 1976) and a refinery in Britain (Hill, 1971). Those involved, managers and workers alike, were selected populations and were volunteers. Projects of this type take advantage of privileged circumstances to demonstrate the reality of certain forays into the future which would otherwise remain no more than untested possibilities. They represent the fullest embodiment of the new model so far attained, in view of the different levels and interest groups involved. Modeling in the new direction is a complex process but has been experienced as easier in new than in already established plants.

In established plants one has to deal with those already there, among whom are those who don't want to change or whose limitations of ability or forms of character prevent them from changing. The accumulated practices of the past are present along with an array of vested interests. If the plant is unionized there will be fear on management's part of surrendering prerogatives and on the workers' of compromising the union's independence. There has to be some sharing of power. Sharing of power is a basic principle in the new model.

In established plants, progress is at best slow, at the worst the change effort has to be abandoned. New methods of process consultation are required. Ketchum (1975) has evolved a practice of uncovering what Argyris and Schon (1974) would call "theories in use" as distinct from "espoused theories." He attempts to unprogram key participants from deep, implicit attachments to the "traditional system" before anything new is proposed. But to cover a whole organizational population in this manner poses problems as yet far from solved in change efforts with social aggregates. Yet whole organizational populations are what one must deal with at this system level.

Continuous Adaptive Planning at Corporate and National Levels

At the corporate level any thoroughgoing change toward the new model would appear to involve the working out of an explicit management philosophy consonant with the new direction. Objectives and policies may then require considerable redefinition. A pioneer attempt in Shell, U.K. (Hill, 1971) involved the participation and collaboration of all groups from the board to the shop floor. Ackoff (1974) refers to a process called "idealized design," through which a corporation can enlarge its conception of what it ought to become by temporarily removing the usual constraints. The "search conference" under development by the two Emerys (1976) involves scanning first the contextual and then the transactional environment, in a future as well as a present perspective. The objective is to develop a means of providing the organization with a wider framework within which to find a new path. These methodologies are all related to what Ackoff (1974) has called continuous adaptive planning. They have been used with communities and in the public sector as well as with industrial organizations. They deal with what Parsons (1960) called the institutional level of organizational life at which questions of value critical for the new model are paramount. They require the participation and collaboration of all interest groups.

Some experience is available at the industry level. Thorsrud and his colleagues have for some years been engaged in a collaborative project with the Norwegian Shipping Industry (Herbst, 1976). This comprises a system of "organizational ecology" (Trist, 1976). Though all the organizations belonging to the system are closely interdependent, no single one is in overall control. If the old bureaucratic pattern were followed, the danger is that a form of corporatism might emerge that would lead in a totalitarian direction. The new institution-building task is to discover an alternative route based on participative and democratic principles which can secure interorganizational collaboration.

In Norway, also, there is evidence of collaboration's working at the national level. The Norwegian Industrial Democracy Project (Emery & Thorsrud, 1964; 1976) has from its beginning in the early 60s been actively sponsored by the Confederation of Employers and the Confederation of Trade Unions. At times the Government has also joined in. A small country with a homogeneous population and culture, with democratic traditions, and a stable system of industrial relations would seem to offer the best opportunity for pioneer collaborative efforts at this level. A critical learning arising from this whole project concerns what might be called "a new culture of politics" (Trist, 1970), the essence of which seems to be a process of collaboration which transcends negotiation. In the case under discussion major conflict was avoided which would have further disturbed the country as a whole at a time when turbulence was already rising from several sources. A new culture of politics will certainly be required if the new organizational model is to be effective at the societal level.

A STEP FORWARD

The purpose of this survey has been to show that types of collaborative process, consonant with the need to fashion a new organizational paradigm capable of absorbing and reducing environmental turbulence, have begun to take place in work settings at all system levels. In the last five years they have been occurring more rapidly and more extensively than in the last ten. Though more evident in the industrial sector, recently there have been some signs of their appearance in work organizations in the nonindustrial sector—in the civil service, in health and education, and in some community endeavors. Most Western countries have now managed to get some collaborative programs under way, and a very few developing countries (India, Peru, Mexico) have made fragile beginnings. Eastern European countries remain outside the trend, unless Yugoslavia is counted; but the Yugoslav Workers Councils are apparently as yet rather autocratic. From afar the Chinese communes look promising but remain enigmatic.

What has been so far accomplished may seem small in face of the vast task that remains to be done. But it is a step forward to have a full view of the new collaborative path and to know how critically important it is to proceed along it in all work settings.

REFERENCES

Ackoff, R. L. *Redesigning the future*. New York: Wiley, 1974.

Ackoff, R. L., & Emery, F. E. *On purposeful systems*. Chicago: Aldine-Atherton, 1972.

Allport, G. *Personality: A psychological interpretation*. New York: Henry Holt, 1937.

Argyris, C., & Schon, D. *Theory in practice*. San Francisco: Jossey-Bass, 1974.

Ashby, W. R. *Design for a brain*, 2nd ed. New York: Wiley, 1960.

Bennis, W. *The crisis of transition in organizational frontiers and human values*. W. H. Schmidt, ed. Belmont, Calif: Wadsworth, 1970.
Chein, I. The environment as a determinant of behavior. *Journal of Social Psychology*, 1954, **39**, 115-127.
Emery, F. E. The next thirty years: Concepts, methods and anticipations. *Human Relations*, 1967, **20**, 199-237.
Emery, F. E. *Futures we're in*. Canberra: Centre for Continuing Education, Australian National University, 1974. Leiden, The Netherlands: Martinus Nijhoff, Social Sciences Division, 1977.
Emery, F. E., & Emery, M. *A choice of futures*. Leiden, The Netherlands: Martinus Nijhoff, Social Sciences Division, 1976.
Emery, F. E., & Thorsrud, E. *The form and content of industrial democracy*. London: Tavistock Publications, 1964.
Emery, F. E., & Thorsrud, E. *Democracy at work*. Leiden, The Netherlands: Martinus Nijhoff, Social Sciences Division, 1976.
Emery, F. E., & Trist, E. L. The causal texture of organizational environments. *Human Relations*, 1965, **18**, 21-31.
Emery, F. E., & Trist, E. L. *Towards a social ecology*. New York: Plenum Press, 1973.
Friedmann, J. *Retracking America: A theory of transactive planning*. New York: Doubleday Anchor, 1973.
Galbraith, J. K. *The new industrial state*. New York: Houghton Mifflin, 1967.
Goldstein, K. *The organism*. New York: American Book Co., 1939.
Herbst, P. G. *Alternatives to hierarchies*. Leiden, The Netherlands: Martinus Nijhoff, Social Sciences Division, 1976.
Herzberg, F., Mausner, B., & Synderman, B. *Motivation to work*. New York: Wiley, 1959.
Hill, C. P. *Towards a new philosophy of management*. London: Gower Press, 1971.
Ketchum, L. A case study of diffusion. In L. E. Davis and A. B. Cherns (Eds). *The quality of working life*, Vol 2. New York: Free Press, 1975.
Maccoby, M. *The gamesman*. New York: Simon & Schuster, 1976.
Michael, D. N. *On learning to plan—and planning to learn*. San Francisco: Jossey-Bass, 1973.
Ozbekhan, H. Planning and human action. In P. A. Weiss (Ed.), *Hierarchically organized systems in theory and practice*. New York: Hafner, 1971.
Parsons, T. *Structure and process in modern societies*. Glencoe, Ill.: The Free Press, 1960.
Perlmutter, H. *Towards a theory and practice of social architecture*. London: Tavistock Publications, 1965.
Schon, D. *Beyond the stable state*. London: Temple Smith, 1971.
Selznick, P. *Leadership in administration*. Evanston, Ill.: Row and Peterson, 1957.
Thompson, J. D. *Organizations in action*. New York: McGraw-Hill, 1967.
Trist, E. L. Urban North America—the challenge of the next thirty years. *Plan Journal of the Town Planning Institute of Canada*, 1970, **10**, 3-20.
Trist, E. L. A socio-technical critique of scientific management. In D. O. Edge and J. N. Wolfe (Eds.) *Meaning and control*. London: Tavistock Publications, 1973.
Trist, E. L. Action research and adaptive planning. In A. W. Clark (Ed.), *Experimenting with organizational life*. New York: Plenum Press, 1975.

Trist, E. L. A concept of organizational ecology. *Bulletin of the National Labour Institute* (New Delhi), 1976, **12**, 483-496.

Trist, E. L., & Bamforth, K. W. Some social and psychological consequences of the longwall method of coal getting. *Human Relations*, 1951, **4**, 3-38.

Trist, E. L., Higgin, G. W., Murray, H., & Pollock, A. B. *Organizational choice: Capabilities of groups at the coal face under changing technologies*. London: Tavistock Publications, 1963.

Vickers, Sir Geoffrey. *The art of judgement*. London: Chapman & Hall, 1965.

Walker, C. R., & Guest, R. H. *The man on the assembly line*. Cambridge, Mass.: Harvard University Press, 1953.

The Assembly Line: Its Logic and Our Future

F. E. Emery

Henry Ford's introduction of the continuous flow conveyor to assembly of automobiles was the apogee of a modern mode of human management—a mode that started to emerge with Frederick Taylor and Gilbreth in the late 19th century. General Motors' much publicised fiasco with their Lordstown plant in 1972 seemed to mark the end of that era. Volvo's announcement, within the following months, of their radically new concept for their Kalmar assembly plant in Sweden seemed to announce the beginning of a new era of human management.

The logic of human management that was enshrined by the car makers became the logic that was worshipped by practically all other large scale manufacturers. It came to be seen in the forties and fifties as the only *sure* way of manufacturing increasingly complex products at a cost price that could serve a mass market. As the real costs of labour have increased through the 20th century it has been seen as the only *possible* way of creating mass markets, or of meeting the demands these mass markets made for industrial inputs. So, it was not too surprising to find a process industry like ICI (U.K.) as the leading proponent in Britain in the post-war years of the philosophy of Henry Ford.

The logic of the car assembly line is probably *the* keystone to prevailing 20th century concepts of human management. The prevailing climate of opinion, however, enables us to attempt to pull out this keystone.

Reprinted from *The Journal of The Institution of Engineers, Australia*, May-June, 1975, pp. 11-13, 16. Used with permission.

The logic of production by a continuous flow line was well understood in the early phases of industrialisation. First Charles Babbage and then Karl Marx spelt this out. The logic was an 'if X then Y' logic. If a complex production task was broken down into a set of constituent tasks then the level of craft skills required was lowered and hence the cost of labour was lowered. At the extreme, a class of unskilled labour emerged to perform the very elementary tasks that practically always remained after a complex task has been broken down to its minimum skill requirements. Such unskilled labourers would have had no part to play in craft production. There is another valuable side of this penny that was not obvious till a much later date. That is, that if one lowers the level of craft skill needed for a product, it becomes much easier to swing to production of major variations of that product e.g. from swords to ploughshares. Massive re-training of craftsmen is not needed.

Once the partition of a task had been successful in reducing the necessary level of craft skills it was only natural that men should seek further partitions leading them into even broader and cheaper labour markets. World War II gave a further great stimulus to this approach. The military demanded large-scale production of very complex machines when often the only available labour force was that conscripted from outside the traditional industrial work force, e.g. women, pensioners and peasants. The lessons flowed over from blue collar work to the organisation of offices—insurance offices, taxation etc., organised for mass-flow production documents.

COST FACTORS OF THE ASSEMBLY LINE

However, to realise the economic advantages of task segmentation it was necessary to cope with several sources of cost *inherent* in the method—transfer costs, standardisation, 'balancing', external co-ordination and 'pacing'.

Transfer Costs

Individual craft production requires a minimal movement of the object under production. Partitioning of production requires transport of the object between each of the work stations at which someone is performing a different sub-task. The costs are those of sheer physical movement and re-positioning so that the next operation can be proceeded with, and also, the costs of 'waiting time' when valuable 'semi-products' are, as it were, simply in storage. Henry Ford's introduction of conveyor belts to car assembly seemed to be the natural outcome of the attempt to reduce these transport costs. Conveyor chains had already transformed the Chicago slaughter houses. Palletisation and fork-lift trucks continue to reduce these costs in assembly areas where continuous belts or chains are not justifiable.

Standardisation

Partitioning of a production process was simply not an economic proposition unless there was a fair probability that the separately produced parts could be re-constituted to yield a workable version of the final product (it would not have to be as good as the craft-produced product if its cost was sufficiently lower). This was obvious enough with the 18th century flow-line production of pulley blocks at the Woolwich Naval Arsenal and the mid-19th century production of rifles. Reduction of this inherent cost has a long history. From Mandsley's slide rest onwards there has been a continuous evolution in specialised tools, machines, jigs and fixtures to enable relatively unskilled labour to continuously replicate relatively skilled operations to a higher degree of standardisation. The most radical developments emerged in the second quarter of this century with metrology, the sophisticated concept of tolerance levels expressed in statistical quality control and national standards authorities. The difficult emergence of this latter revolution in Australian industry is well documented in Chapter 7 of Mellor's volume of the World War II History (Ref. 5).

'Balancing the Line'

This problem does not arise with the individual craftsman. Whatever the problems with a given phase of production on a particular lot of raw materials he can proceed immediately to the next phase as soon as he is satisfied with what he has done. He does not have to wait to catch up with himself. When the task is partitioned that is not possible. Each set of workers is skilled, or rather, semi-skilled, only in their sub-task. They are not skilled to help clear any bottleneck or make up any shortfall in other parts of the line; they can simply stand idle and wait. Theoretically there is in a flow-line an 'iron law' of proportionality, such as Marx writes of, which should be like the recipe of a cake: so many hours of this kind of labour, so many of that kind of labour etc., and hey presto, the final product. Unfortunately for the application of the theory it is not as simple as making cakes.

Balancing the line, to reduce downtime, was an on-the-line art of observation until Taylor and Gilbreth came on to the scene. Their contribution was Methods-Time Measurement (M.T.M.). At last the balancing of a line seemed to be a science. Controlled observation and measurements seemed to offer a way of balancing not just the labour requirements of the major segments of a line, but of scientifically planning the work load and skill level of each and every individual work station. Planning and measuring costs money, but there has seemed no other way to reduce the downtime losses inherent in the original fractionation of production.

External Supervision and 'Pacing'

So long as the individual craftsman produced the whole product, control and coordination of his work on the various sub-tasks was no problem. He managed that himself. With the fractionation of production a special class of work emerged, the work of supervision. Each person on the assembly line has to attend to his own piece of the work and hence someone else must co-ordinate what is happening at the different work stations, to re-allocate work when the line becomes unbalanced, and to re-enforce work standards when individual performance drifts away from them. A major headache has been the near universal tendency of workers on fractionated tasks to drift away from planned work times. The self-pacing that enables the craftsman-producer to vary his work pace and yet maintain good targets for overall production times appears to be absent from small, fractionated tasks that are repeated endlessly. Tighter supervision and incentive payment schemes seemed appropriate forms of the caveat and stick to replace this element of self-pacing. However the moving line emerged as the major innovation.

Once properly manned for a given speed it seemed that this speed had only to be maintained by the supervisors to ensure that planned work times would be maintained. Dawdling at any work station would quickly reveal itself in persons moving off station to try and finish their parts.

However, it is not quite as simple as that, for it is possible on some work stations to let unfinished work go down the line with a chance of it not being detected until the product is in the consumer's hands. Anyway, the main point was readily learnt. The conveyor was not just a means of lowering transfer costs but also of reducing supervisory costs. At certain tempos the line even gave operators a satisfactory sense of work rhythm; a feeling of being drawn along by the work. Davis' 1966 study (Ref. 2) even suggests that the contribution to control may often be the main justification of the conveyor.

VOLVO AT KALMAR: SEEKING THE LOGICAL LIMITS

What I have spelt out is old hat to any production engineer. Nevertheless, it prepares the ground for the point of this paper.

We have heard a great deal lately about the demise of the automobile assembly line. The new Volvo plant at Kalmar does not even have conveyor belts. The EEC pronounced in 1973 that the assembly line would have to be abolished from the European car industry.

I suggest that the new Kalmar plant does not represent any departure from the basic principles of mass-flow line production. In the first place they are still seeking the maximum economic advantages to be gained from fractionation of the overall task. In the second place the plant and its organisation have been designed to reduce the same inherent costs of mass flow production, i.e. costs of transfer, standardisation, balancing coordination and pacing. Kalmar is designed as a mass production flow line to produce an economically competitive product. There is no radical departure from the principles of flow-line production, only from its practice.

Note, however, that I previously stressed that their aim was 'maximum economic advantage from fractionation' *not* maximum fractionation.

What they have done is to recognise that the costs we have discussed are inherent in production based on fractionated tasks. The further one pushes fractionation the greater these costs become, particularly the costs other than transfer because they are more related to human responsiveness. *The objective of gaining maximum economic advantage from fractionation cannot be the objective of maximum fractionation*. There is some *optimal* level to be sought at a point before the gains are whittled away by rising costs.

THE TRADITIONAL ASSUMPTIONS

Why has this logic been so obvious to the Kalmar designers and yet appeared to escape other car plant designers? I do not think it is just because M.T.M., quality control, production supervision, designers etc. operate in separate boxes with their own departmental goals. After all, at the plant design phase there are usually opportunities for the various specialties to come together. I think the reason lies deeper. If we look at the traditional practices in designing a mass-flow line we find a critical assumption has slipped in and been reinforced by the widespread reliance on M.T.M., as a planning tool and as a control tool. This assumption is that it must be possible for each individual worker to be held responsible by an external supervisor for his individual performance. On this assumption M.T.M. goes beyond being a planning tool to determine or re-determine the probable labour requirements of sections of the line. It becomes part of the detailed day-to-day supervisory control over production. Under this impetus fractionation heads down to the L.C.D. (Lowest Common Denominator) of the labour on the line.

The same assumption that a line must be built from the individually supervised one-man-shift unit has gone into the design of algorithms to determine line balance. Ingall (Ref. 3) has reviewed ten or so of the major algorithms which all embody the same assumption. They go further along with M.T.M. to assert that this is a firm organisational building block by assuming that, on average, different operators work at the same pace, on average, an individual operator works at the same pace throughout a shift, on average, cycle time of the operation is irrelevant, on average, learning on the job can be ignored, on average, variations in parts and equipment can be ignored.

An average is just that, an average. It represents the mean value of a set of different observed states of a system parameter. It does not even tell us whether the exact average state has ever occurred. One thing is pretty sure: at any one time on a line it is most improbable that all aspects are operating at their average value. Typically something is always non-average, wrong, and when one thing is wrong so are half a dozen other things.

The practical problems of balancing a line simply cannot be solved by abstracting this aspect from the total system of potential gains and inherent costs of flow production. As Ingall concludes in his review of assembly line balancing:

> Knowing whether these problems occur together is important because analysing them separately is not sufficient if they do. Using the "sum" of the results obtained by analysing each problem separately as the procedure for the combined problem can be a dangerous pastime (Ref. 3).

The practical significance of the balancing problem may be gauged from the finding of Kilbridge and Wester (Ref. 4) that the U.S. Automobile industry wasted about 25 per cent of assembly workers' time through uneven work assignment. No doubt this figure had been reduced at Lordstown in 1972.

I have wandered a little afield because I wish to stress how far this assumption about the individual building block has unquestionably grown into the professional way of looking at the assembly line.

It is this, I suggest, which has prevented others from seeing the obvious logic of the line as did the Kalmar designers.

This hidden assumption has, I think, had a further distorting effect on thinking about the line.

Some people in the car industry during the fifties and the sixties became sensitive to the fact that pursuit of maximum fractionation was self-defeating; and they realised it was not at all like the engineering problem of pursuing maximum aircraft speed by reducing friction and drag. It was not a problem to be solved by the grease of yet higher relative pay, by featherbedding or by any of those things that Walter Reuther of the U.S. Automobile Workers' Union bitterly referred to as 'gold plating the sweat-shop'.

The response of these people to such critical insights was to look again at the building block, the one-man-shift unit, to see what could be done about that. They did not question whether the individual was the appropriate block for building on.

One proposal to arise from this was to employ on the line only people who were at or very close to the lowest common denominator used by M.T.M. and the planners, i.e. donkeys for donkey work. This proposal does not look so good now that the international pool of cheap migrant labour dries up. In any case there was little future in this proposal. Provided the line designers pursue the same twisted logic of maximum fractionation, they would inevitably design around an even lower common denominator and costs would rise again.

The other proposal was to discard the concept of an L.C.D. and accept job enlargement or enrichment up to a point which would come closest to optimal fractionation for a majority of the people on the line. Imposing such enrichment on the minority whose optimal was below this level was an immediately obvious practical flaw. A more deep-seated flaw was that this job enrichment approach argued from consideration of only one aspect of the system—task fractionation. It did not simultaneously confront the other parameters of the system—balancing, pacing etc.

Here we come to what is really radical about the Kalmar design.

The designers approached their task with an awareness that the problems of flow-line production could be theoretically approached in different organisational designs. At one extreme they literally examined the old cottage industry. More seriously they compared the Norwegian experiences (Ref. 7) with the semi-autonomous work group as the building unit as against the traditional individual-shift-work-station unit.

THE SEMI-AUTONOMOUS WORK GROUP

The most striking outcome was the discovery that in an appropriately skilled and sized work group all of the key parameters of mass flow production could come together and be controlled vis-a-vis each other at that level. Picturesquely this was labelled as 'a lot of little factories within a factory'. In terms of how we picture a factory, this the groups are not. Walk around Kalmar and you see nothing that even looks like a lot of little workshops producing their own cars. In system terms, however, it is a very apt description, a very valid design criterion.

This becomes more apparent if we look at the production groups formed at Kalmar.

The first effects are rather like those that the northern N.S.W. pig farmers have gained from co-operation. Individually their resources gave them little or no freedom of movement and they had to ride with a market that was basically out of their control. Collectively they have found new degrees of freedom and they have started to shape their markets to allow even more freedom.

Formation of semi-autonomous groups on the Kalmar assembly line has given them a cycle time and buffers that would be negligible if split up for individual work stations. Split up into individual work stations no-one could take an untimed coffee break; grouped up everyone can without increasing overall downtime. On individual work stations everyone has to meet the standard work on that particular job minute by minute. In the group-setting variations in individual levels of optimal performance can be met hour by hour. Those that prefer repetitious simple tasks can get them: those that need to be told what to do will be told by others in the group. Within the range of their task the group can balance their work, without outside assistance. If quality control is amongst the group's responsibilities, and they are given time allowance for this, then it can be within their capabilities. Now we come to the fundamental matter—co-ordination, control and pacing.*

If a semi-autonomous work group is not willing to exercise control and co-ordination over its members then the design of flow-lines must go back to the traditional model.

At this point I must rely on experience not theory. The experience, over twenty odd years with a wide range of technologies and societies, is simply this: *if reasonably sized groups have accepted a set of production targets and have the resources to pursue it at reasonable reward to themselves, they will better achieve those targets than they would if each person was under external supervisory control.* If a theory is required then I think it need be no more than that that is spelt out in the six psychological requirements defined in 1963 (Ref. 6). In groups that have sufficient autonomy and are sufficiently small to allow face-to-face learning these criteria can be maximally recognised. It has been the

---

*Volvo engineers came up with an ingenious technical solution to the transfer problems—the individual self-propelled carriage. This allows the groups to vary the times put into each car while maintaining average flow on to the next group.

realisation of these individual human requirements that has enabled semi-autonomous group working of mass flow lines to do what could not be done by M.T.M. and algorithms for balancing the lines.

CONCLUSION

The revolution at Kalmar has not been that of throwing out the assembly line. The revolutionary change began with the eradication of an organisational principle of one man, one shift, one station—a principle that had no intrinsic relation to the design of assembly lines.

REFERENCES

1. BASU, R. N.—The Practical Problems of Assembly Line Balancing. *The Production Engineer*, Oct. 1973, pp. 369-370.
2. DAVIS, L. E.—Pacing Effects on Manned Assembly Lines. *International J. of Industrial Engineering*, 1966, Vol. 4, p. 171.
3. INGALL, E. J.—A Review of Assembly Line Balancing. *J. of Industrial Engineering*, Vol. 16, No. 4, 1965.
4. KILBRIDGE, M. D. and WESTER, L.—The Assembly Line Model-Mix Sequencing Problem. *Proc. of 3rd International Conference on Operational Research*. Oslo, 1963.
5. MELLOR, D. P.—*The Role of Science and Industry*. Canberra, Australian War Memorial, 1958.
6. THORSRUD, E. and EMERY, F. E.—*Form and Content in Industrial Democracy*. Oslo, University Press, 1963; London, Tavistock, 1969.
7. THORSRUD, E. and EMERY, F. E.—*New Forms of Work Organisation*. Oslo, Tannum, 1969 (in Norwegian). Revised English edition: EMERY, F. E. and THORSRUD, E.—*Democracy at Work*. Canberra, Centre for Continuing Education, ANU, 1975.
8. WILD, R.—Group Working in Mass Production, Part 1: Flowline work. *The Production Engineer*, Dec. 1973, pp. 457-461.

Social Relationships and Organizational Performance: A Sociotask Approach

William A. Pasmore

Suresh Srivastva

John J. Sherwood

In this paper, we are concerned with relationships among people and groups of people performing tasks in organizational settings. Some time ago, Barnard (1938) noted that organizations are formed to accomplish tasks that individuals cannot accomplish working separately; he concluded that organizations are by nature cooperative systems and cannot fail to be so. Given this view, the performance of an organization is largely determined by the ability of its parts to perform their interdependent tasks, i.e., to the extent that people and groups in organizations do in fact accomplish those tasks that they could not accomplish working alone, the organization will be successful. Conversely, to the extent that people and groups in organizations concentrate exclusively on the performance of their individual tasks and ignore cooperative efforts, the performance of the organization will be hindered.

Our observations and those of others lead us to believe that most organizations are not nearly as cooperative as they could or should be. Instead, most organization members are trained to do certain specialized tasks, they are evaluated on the basis of individual task accomplishment, and they are individually selected and promoted. As a result, organizations often lack the cooperative element needed to achieve a synergistic organizational effect and concomitant success.

Most organizations are designed as collections of individual tasks, allocated according to the demands of the technology, structure, or needs of the organization at a given point in time. Although these arrangements may work efficiently for a period of time, the inflexibility of individual task assignments weakens the organization's ability to respond to change. Market trends indicating the need to change products may be overlooked, or governmental legislation that would change work patterns might be ignored. Until actual crises occur, individual tasks continue to be performed in a mechanical, oblivious manner.

When a crisis does occur, managers and other leaders are called on to fill the vacuums created by new internal or external demands on the organization. They are asked to "make certain that everyone is working together toward a common goal" or to "better utilize existing resources to meet new challenges." This task is more difficult than most suspect, and to date the behavioral sciences research has not offered much assistance.

For these reasons, our focus in the paper is on organizational tasks that require the cooperation of individuals or groups. Tasks, for the purpose of this paper, are activities performed by individuals or groups in organizations that lead to the accomplishment of organizational goals. We assume that all tasks in an organization are interdependent to some extent, i.e., because organizations are cooperative systems designed to perform tasks that individuals could not accomplish working alone, the performance of every task necessarily affects the performance of every other task within the organization. Our concern is those tasks which, if *not* performed cooperatively and in a coordinated manner, will have a significant impact on organizational performance. These task relationships are referred to as "key" because the quality of their performance critically affects organizational effectiveness.

The notion of interdependence was clearly explicated by John C. Glidewell (1970), who provided some crucial insights into the interdependence of individuals and groups within organizations. Although Barnard recognized that organizations are by nature cooperative systems, his idyllic notions of cooperation by employees' "organizational personalities" have been seriously questioned (Perrow, 1972). Not long after Barnard wrote *The Functions of the Executive* (1938), Weber's classic accounts of bureaucracy (almost the antithesis of Barnard's ideas) were translated (Weber, 1946, 1947). More recently, scholars have pointed out that organizations cannot rely on their members to carry out assignments voluntarily and have emphasized the importance of control and authority systems (Etzioni, 1965; Galbraith, 1977).

In contrast, Glidewell (1970) suggested that cooperation in organizations can be achieved if it is the product of commitments among individuals and groups to perform certain activities on a regular and dependable basis in exchange for some equitable distribution of resources that otherwise could not be obtained. He stated:

> Social organization requires commitments to others, and from others. Each man puts into the organization some motives, feelings, ideas and skills. He gets out of the organization other motives, feelings, ideas and skills, at the times and in the places and in the manner they can be used. The time and place and manner have to be reasonably certain, and that requires commitments to others and from others. It requires dependability. (p. 56)

According to this view, cooperation in organizations is the result of a series of exchanges between individuals or groups. *If* these exchanges take place between the appropriate parties, and *if* the agreements reached are perceived to be equitable, a cooperative system will develop. Unfortunately, in most organizations such exchanges occur more by chance than by plan; the result is informal, hit-or-miss cooperation for performance of key interdependent tasks. To the extent that the need for cooperation on key tasks can be recognized and procedures can be developed to facilitate dependable performance agreements, the effectiveness of an organization should improve.

Sociotechnical system interventions have achieved some success in designing blue-collar production processes to deal with interdependent tasks (Srivastva et al., 1975). Through the formation of autonomous work groups (Bucklow, 1966) employees are able jointly to accomplish tasks that have traditionally been assigned to individuals, despite their key impact on the performance of significant organizational units.

Too little attention has been paid in the past to designing systems for the performance of key interdependent tasks that do not utilize well-defined technologies, such as hospitals, universities, political organizations, and other service-oriented enterprises staffed by white-collar personnel. Also, the roles of management and other support groups have been overlooked too often in the redesign of organizations, despite their influence on key task interdependencies. A new and more comprehensive theory is needed specifically to address social and task-system designs in all organizations and at all organizational levels. To distinguish this more general theory from its predecessors, we call it the theory of *sociotask design*.

Sociotask theory can be outlined in the form of propositions, noting how the design of organizations as sociotask systems may be affected by the nature of tasks performed, the technology employed, the organizational structure, and the power relationships among individuals and groups.

DEPENDABILITY AND THE NATURE OF TASKS

An equitable exchange is fundamental to the formation of effective, enduring, and satisfying relationships. Upon joining an organization, individuals learn that certain tasks are required of them. Some of these tasks may seem more important than others; some may be carried out alone, and others may require the assistance of other individuals or groups. In the early stages of the individual-organization relationship, each new member weighs the benefits derived from performing assigned tasks in terms of the monetary, social, and other rewards received. If these benefits outweigh or balance the amount of effort the individual must put forth and if the individual is unaware of other more satisfactory jobs elsewhere, he or she becomes socialized and remains with the organization. In this way, a social contract (unstated and unwritten) forms between the individual and the organization specifying that the employee will perform the tasks required in return for the rewards offered. If the employee feels that the contract is equitable, he or she will enter into the contract and become a dependable member of the organization. If the rewards experienced are not greater than or equal to the effort that must be expended in task performance, or if alternatives are judged to be more desirable, the individual will not enter into the contract and will not become a dependable member of the organization. This view of the socialization process has been stated in various ways by Barnard (1938), Simon (1957), March and Simon (1958), Thibaut and Kelley (1959), and Thompson (1967).

Of course, the measurement of rewards experienced and effort expended is very idiosyncratic; what is equitable for one individual may not be tolerable for another. Nevertheless, the judgments that are made are considered by those affected to be relatively stable. If at any future time a change is made in the tasks to be performed or the rewards to be earned, the individual must renegotiate the social contract and either remain and assume the new relationship or seek employment elsewhere.

Over time, as the individual's relationship with the organization develops, he or she will learn that certain of the tasks originally assigned are essential and must be executed before the organization will fulfill its part of the bargain. These tasks are *prescribed*, i.e., the organization specifies that their performance is absolutely necessary and that they must be performed in a certain way. Other tasks may be overlooked without disturbing the relationship. These tasks are *discretionary*, i.e., the individual may exercise some choice in determining their frequency or method of performance.

The individual also will make agreements with certain other individuals or groups to perform some activities that will presumably benefit both parties. Hence, a contract of sorts is formed between the parties; both expect to perform and be rewarded for tasks not originally specified by the organization. Because these tasks require the establishment of reciprocal agreements, they are referred to as *contractual*.

Finally, the employee may find that certain tasks are required to maintain the viability of the organization in its environment, although these tasks may not be specified by the organization. Relationships must be maintained with the environment to secure materials, provide outlets for goods and services, and respond to legislation. Because these tasks spring from demands made on the organization by the environment, they are referred to as *emergent*. A detailed analysis of the four task categories is provided by Cummings and Srivastva (1977).

The terms "other individuals" or "other groups" could be substituted for "organization." In much the same way that relationships are formed between the individual and the organization or groups within the organization, relationships are formed among groups or departments that make up the organization. These relationships also involve prescribed, discretionary, contractual, and emergent tasks. Groups that are forced to work closely together by the nature of the technology, as on the assembly line, have a prescribed relationship; union and management groups meet jointly to discuss discretionary tasks such as planning an annual picnic; research and development and marketing groups may enter contractual understandings about the rate of new additions or about changes in the product line; corporate legal and personnel staffs may work together to deal with Equal Employment Opportunity guidelines or other emergent tasks.

The organization forms relationships with its environment in a similar fashion. Some tasks must be performed by the organization in order to survive, i.e., it must produce some product or perform some service desired by the environment in return for resources; the organization also may provide other products or services not demanded by the environment, such as skill training for disadvantaged youths; they may work with government or other agencies on a contractual basis to research new ideas, which is frequently true for universities; or they may work on solutions to emergent problems such as pollution or international relations.

The key elements in all of these relationships are *equity* and *dependability*. Unless the parties to the relationship experience balanced outcomes in terms of reward for effort, the social contract will not be stable and organizational performance will be hindered because there is little motivation for the party receiving the lesser reward to maintain the relationship. The parties also must be able to depend on one another to provide the goods and services required to fulfill the contract on a regular basis. Sporadic compliance engenders renegotiation of the contract, with the more dependable party demanding greater rewards or offering less effort than in the past.

The nature of the task determines, in part, the reaction of task-related groups to perceived inequities. The performance of prescribed tasks is usually critical to the organization's success, so this performance is monitored and enforced by the organizational control system. If the relationship between two parties is prescribed but perceived as inequitable, the less satisfied of the two will seek some form of control over the other. The individual or group may refer the matter to a superior if there is one, demanding the enforcement of an agreement that they believe to be essential to the organization's well-being. For example, assembly workers may seek the assistance of the production supervisor to secure more dependable delivery of materials from the warehouse or a doctor may call on the hospital administrator to speed up the delivery of important medical records and charts.

As an alternative, the other party may be approached directly, either with demands for compliance or threats to withhold cooperation. This is frequently true in union-management relationships. Regardless of the approach used, if the outcome of the negotiation process is unsuccessful, conflict often develops between the parties. If the relationship involves a key interdependent task, conflict adversely affects organizational performance.

In most organizations, the prescribed tasks and interrelationships are inadequate to ensure proper functioning of internal units. Individuals and groups find ways to take up the slack in terms of performing tasks that are discretionary in nature. In times of organizational stress or change, these tasks are usually the first to be discontinued, often to the further detriment of relationships within the organization. Concentration is given to the primary prescribed tasks, and the result is that critical relationships are not developed (Miller, 1971). This pattern was manifest in the cutback in internal organization development staffs by corporations during the recessions of the early Seventies; organizations applied the resources that had been used to support these staffs to the primary production process instead. Only now are some of these organizations rebuilding their internal consulting groups to respond to social and technological problems that might have been avoided altogether had these staff groups been maintained continuously.

When discretionary tasks are not performed dependably, there is usually no way for others to demand that they be done. When a personnel manager fails to organize the company picnic as expected, others may be dissatisfied but probably will not demand his or her resignation. Similarly, if physician's assistants have assumed the discretionary role of go-between for doctors and nurses, the latter have no power to demand the continued performance of this function *unless* the go-between role is a prescribed part of the position. Usually, changing task performance from discretionary to prescribed involves offering greater compensation or status to those who perform the task or writing the tasks into job descriptions over an extended period of time.

If the relationship between two task-interdependent parties is contractual, the recourse for undependable performance is determined by the nature of the contract. If the contract is written, which is somewhat rare, it will often specify the steps to be taken to settle performance issues equitably. If the contract is simply understood, as is most often the case, the renegotiation process may take a variety of forms. The disadvantaged group may withdraw its support or information from the other, cease to perform its function completely, or, in extreme cases, perform its task in a way that affects the performance of the other group adversely. Once it is obvious to the delinquent group that its counterpart is demanding renegotiation of the contract and that such renegotiation no longer can be avoided, leaders or representatives from the groups will meet to discuss the alternatives. Through negotiation, contractual agreements are made clear and become prescribed and predictable, thereby improving organizational effectiveness.

Finally, if two groups are interdependent for the performance of emergent tasks but their relationship is perceived by one or both to be inequitable, performance of the task again will suffer. Public relations and research and development departments are often called on to work closely together to further the organization's position in its environment. Research and development must keep the public relations department informed, yet not allow the dissemination of design secrets. Unless the relationship between the groups is carefully specified in terms of what information will be available for public scrutiny, the position of the organization in its environment may be adversely affected. The same is true of production planning and marketing groups, legal staffs and personnel departments, and hospital administrators and boards of trustees. In all of these relationships, cooperation and coordination are essential if the organization is to be at all proactive instead of reactive to its surroundings.

Unfortunately, the relationships between groups working on emergent tasks are probably the least specified, due to the unpredictable nature of the environment. Because the day-to-day tasks of these groups are not clear and stable, it is difficult to make contractual agreements. In many cases, the relationship between groups evolves through a series of disputes regarding what actions each should *not* take that would hinder the performance of the mutual task. In a sense, the relationship between the groups becomes prescribed through this process because the number of acceptable actions that can be undertaken is reduced to a predictable and stable set. Inequitable performance of these emergent-prescribed tasks can then be settled by referral to a higher authority or through renegotiation, as with other prescribed interdependent tasks.

The reactions of task-interdependent individuals and groups to inequities in task performance and how these reactions are affected by the nature of the task performed can be summarized in a series of theoretical propositions, although further research is needed to support their validity.

Proposition 1. Organizations may be defined as systems of interdependent tasks. To function effectively, these tasks must be performed on a dependable basis (Barnard, 1938; Katz & Kahn, 1966; Miller, 1971; Thompson, 1967).

Proposition 2. Parties within an organization perform certain tasks dependably when the reward received for their performance is perceived as equitable in terms of effort or other costs associated with their performance (Barnard, 1938; Simon, 1957; March & Simon, 1958; Thompson, 1967).

Proposition 3. If the relationship between two task-interdependent parties is perceived as inequitable due to imbalanced exchanges between them, the more dependable party will seek renegotiation of the performance contract.

Proposition 4. If the task-interdependent parties seek to renegotiate their contract, the renegotiation generally will take the form of the more dependable party asking for greater rewards or offering less effort in task performance.

Proposition 5. The nature of the task in part determines the mechanisms used to settle task-performance negotiations. In all cases, the result of the negotiation process moves the agreement for task performance toward a more formal, prescribed relationship.

Proposition 6. As tasks become more prescribed in nature, their performance becomes more regular and dependable. To the extent that key tasks can become prescribed, the functioning of a system will be improved.

Although this may seem to be a Weberian concept of bureaucratic design with prescribed rules, regulations, and control systems, this is not the case. We agree with Perrow (1972) that rules do not deprive organizational members of their freedom; only by specifying through prescribed agreements the areas of freedom actually available to subordinates is there any opportunity for the use of discretion by organizational members. This is particularly true of managerial, professional, and human service tasks, which are often ambiguous by their very nature. When tasks can be prescribed, their performance can be more closely evaluated and more properly rewarded. As a result, organizational control systems can be geared directly toward task accomplishment; areas for further development and training can be identified; subordinates can be taught the processes of task performance; and interdependent parties can work together more dependably. For all of these reasons and more, *prescribed* tasks are *dependable* tasks. As such, they represent a desirable objective for sociotask design.

Unlike Weber, we do not believe that all tasks in an organization must be carefully prescribed; but we do suggest that tasks that are key to the organization's success and require interdependent effort should be more prescribed than others. We also suggest that the sociotask design of an organization should be the product of interactions between those actually performing the tasks rather than between those at the top of the hierarchy or those responsible for codifying policies. In short, we support the use of negotiated rules, when necessary, to secure the dependable performance of key organizational tasks.

TECHNOLOGY AND DEPENDABILITY

The nature of the technology used by an organization has profound influence on the extent to which certain tasks can be performed in a dependable manner. Many of the tasks required for an organization to run smoothly and effectively are not prescribed but easily could be. The technology used by organization members may either further the clear prescription of tasks, or it may mitigate against it.

Thompson (1967) noted that different technologies produce different types of interdependencies among organizational subparts, which in turn dictate different coordinating responses to mutual tasks. Mass production and process technologies, which he referred to as *long-linked*, almost completely specify the interactions that must occur among groups. *Mediating* technologies, as employed by banks or utility companies, call for adjustment to specific client cases in planned, standardized ways. *Intensive* technologies, like those of a hospital or a custom production industry, require coordination by mutual adjustment of those involved.

Another typology useful for discussing the implications of technology for dependable task performance is that used by Galbraith and others (Galbraith, 1970, 1977; Burack, 1975; Duncan, 1972) based on the degree of certainty or uncertainty inherent in the tasks. Galbraith (1977, p. 36) defines uncertainty as "the difference between the amount of information required to perform the task and the amount of information already possessed by the organization."

In production technologies utilizing sophisticated mechanical equipment, there is relatively little uncertainty about how to perform the tasks required; the organization possesses all the information it needs to set prescribed task definitions. In managerial positions as well as in the provision of health care services, on the other hand, uncertainties about task performance are great. Because the organization lacks complete information about how to perform certain tasks, these tasks tend not to be prescribed, but discretionary, contractual, or emergent, and relationships among organizational members regarding their mutual performance develop accordingly. As more information about the performance of these tasks is obtained, they too tend to become more prescribed. For example, managers and professionals often are given the task of coordinating the efforts of others in the organization; one tool used for this purpose is the yearly operating budget. If there is some uncertainty regarding the proper distribution of organizational funds, the budgeting process will be primarily political and probably suboptimal in outcome. In contrast, if an organization develops a system for measuring the impact of different budgeting decisions on organizational performance, the budgeting process can become more rational and prescribed.

Although the technology employed by an organization places certain constraints on the methods that may be used to coordinate the performance of interdependent tasks, most organizations overlook the range of coordinating devices still available. Members are prone to rely on the informal organization for the coordination of nonprescribed tasks when the formal organization could perform much of the coordination desired. Even in organizations that utilize relatively certain technologies, there may be value in reviewing existing interdependent performance agreements. The sociotechnical system method of intervention has made progress in this area; this is because the method carefully considers the dependencies between people and between groups that are created by the technology. Perhaps the most straightforward example of this is the formation of autonomous work groups according to the boundaries of whole tasks and in response to the need to control key process variances. In the autonomous group, social contracts are developed that make the performance of certain tasks by group members more dependable, regardless of the specific location of group members on a given day. Individuals and groups in the organization can form contracts for the performance of key tasks; often, these tasks include activities that are typically discretionary, contractual, or emergent in nature and are usually performed by managers (Walton, 1972; Krone, 1974). Members are hired and trained by the group, schemes of pay distribution are devised, job assignments are rotated by group agreement, and group members maintain contact with relevant portions of the organization's environment.

In professional roles, the prescription of discretionary, contractual, or emergent tasks is sometimes accomplished through management by objectives. The result of the process of management by objectives is the same as that for sociotechnical system interventions: tasks become better defined and their performance becomes more dependable.

We posit the following additional propositions:

Proposition 7. The performance of managers, professionals, and other members of organizations, as well as the performance of organizations, can be improved by making the ends and means of nonprescribed task performance more explicit.

Proposition 8. Certain technologies encourage the explicit delineation of prescribed tasks; other technologies tend to mitigate against task prescription. More specifically, long-linked and highly certain technologies encourage task prescription; highly uncertain mediating or intensive technologies inhibit task prescription (Thompson, 1967; Galbraith, 1977).

Proposition 9. In situations in which the technology does not encourage task prescription, other more direct methods of achieving prescription of key interdependent tasks must be found if the organization is to function effectively.

In conclusion, the technology used by an organization or its interdependent parts is an important factor in determining the dependability of task performance, but organization structure also plays a role.

STRUCTURE AND DEPENDABILITY

The structure of an organization affects the performance of tasks in two ways. First, it influences the nature of the tasks performed (Burns & Stalker, 1961); second, it influences the bases on which organizational decisions are made (March & Simon, 1958). Although the structure of an organization is determined in part by the technology it uses for performing its tasks (Woodward, 1958), at the same time, organizations utilizing the same technology may have different structures and the structure of an organization may have effects on task performance separate from those of its technology.

Burns and Stalker (1961) delineate two types of organizational structures, *mechanistic* and *organic*. Mechanistic structures incorporate many of the features of typical bureaucracies, including specialized differentiation of tasks, coordination by supervisors, precise definitions of rights and obligations, hierarchical control, primarily vertical communication, and insistence on loyalty to the organization. Organic structures, on the other hand, have many of the characteristics of cooperative systems, including concentration on the contributive nature of tasks, continual redefinition of tasks, commitment beyond individual tasks, a network structure of control, lateral communication, information rather than direction giving, and commitment to the organization's tasks.

In mechanistic structures, tasks tend to be prescribed clearly at the outset; in organic structures, tasks initially tend to be more discretionary, contractual, and emergent in nature. According to Burns and Stalker, the type of structure appropriate for a given organization depends on the complexity of the organization's environment. Certain organizations, because of their routine technologies and stable environments, are suited to organic structures and nonprescribed tasks. Hence, the environment of the organization limits the choice of which structure will be utilized, which in turn influences the dependability of task performance.

In mechanistic organizations, the relationships among groups are well established through task prescriptions. In organic organizations, tasks are usually not clearly prescribed, and the coordination of organizational activities is more difficult. The performance of tasks, particularly those requiring interdependent effort, are therefore not as dependable. Task performance *can* be made more dependable over time with effort; nonprescribed tasks can be made more prescribed by establishing equitable performance agreements. Thus, the structure of an organization, like its technology, sets the stage for certain types of organizational relationships but these do not remain unchanged over time.

Organizational structures also affect the dependability of task performance by affecting the ways in which organizational decisions are made. According to March and Simon (1958), organizational structures similar to those described as mechanistic by Burns and Stalker allow members to make *programmed* decisions, e.g., in mechanistic organizations the rules for decision making can be clearly specified and the alternative solutions limited to a familiar and accepted set. Organic structures, on the other hand, do not allow members to make programmed decisions easily, so the organization must rely on the *satisficing* behavior of individual decision makers to achieve its goals. March and Simon argued that, to make organizations more rational in terms of goal accomplishment, the bases on which individuals make decisions should be changed. Perrow (1972) summarized this argument as follows:

> Involved are such things as uncertainty absorption, organizational vocabularies, programmed tasks, procedural and substantive programs, standardization of raw materials, frequency of communication channels usage, interdependencies of units and programs. Such mechanisms affect organizational behavior in the following ways: they limit information content and flow, thus controlling the premises available for decisions; they set up expectations so as to highlight some aspects of the situation and play down others; they limit the search for alternatives when problems are confronted, thus insuring more predictable and consistent solutions; they indicate the threshold levels as to when a danger signal is being emitted (thus reducing the occasions for decision making and promoting satisficing rather than optimizing behavior); they achieve coordination of effort by selecting certain kinds of work techniques and schedules. (pp. 156-157)

In summary, the structure of an organization affects the premises on which decisions are made and the extent to which they can be programmed, thereby affecting the dependability of task performance, and, by *changing* the structure of an organization, the techniques of organizational decision making can be improved, with resulting increases in the dependability of interdependent task performance. The following propositions deal with organizational performance and structure:

Proposition 10. Mechanistic structures encourage task prescription, while organic structures inhibit the explicit prescription of tasks (Burns & Stalker, 1961).

Proposition 11. To the extent that tasks in organic organizations can be made more explicit and prescribed, task performance and organization performance become more dependable.

Proposition 12. The structure of an organization affects the way in which individuals and groups reach decisions and thereby influences the level of dependability of task performance (March & Simon, 1958).

Proposition 13. To the extent that the structure of an organization can be utilized to make decision making by organization members more programmable, task performance becomes more dependable and organizational performance is likely to improve.

To portray organizational behavior as determined strictly by the tasks performed, technology utilized, and organizational structure would be less than realistic, however. Consideration also must be given to the differences in power among those who make up the parts of the organization.

POWER AND DEPENDABILITY

From the time that an individual commits himself or herself to a contract for dependable performance in an organization, he or she has a certain amount of power to use in shaping agreements about interdependent task performance. Each member of an organization can determine, within some limits, the extent of cooperation offered in the performance of tasks. As organizations grow and become differentiated, certain individuals acquire more power to affect performance contracts than others. According to Emerson (1962), the power of individuals and groups in organizations is a function of the organization's need for the resources provided by the individual or group and the ability of the organization to acquire the same resources elsewhere. Using this notion of power, Thompson (1967) defined certain cooperative strategies that may be used by organizations or their members to gain power.

He stated:

> Using cooperation to gain power with respect to some element of the task environment, the organization must demonstrate its *capacity to reduce uncertainty* for that element, and must *make a commitment* to exchange that capacity. (p. 34; italics in original)

When negotiating an equitable and dependable performance contract, one party's power is determined both by his or her ability to perform dependably *and* by ability to help the other party perform his or her task dependably as well. Not surprisingly, power is greatly influenced by the nature of the tasks being performed.

Persons performing prescribed tasks hold relatively little power to change agreements about interdependent task performance. Rather, the manager or professional performing nonprescribed tasks is typically afforded greater power in organizations; these individuals have the freedom and the resources at their disposal to facilitate dependable task performance. It is not surprising that efforts to enrich jobs at one level often have a negative impact on interpersonal relationships (Lawler, Hackman, & Kaufman, 1973; Alderfer, 1969); when supervisors, who hold the power to determine the conditions of performance agreements, are not involved in the job-redesign process, no other result can be expected. This has led to a regrettable split between supervisory and other groups.

If an organization's purpose is to offset forces that undermine human cooperation (Scott, 1961), then the redistribution of power to allow the dependable performance of interdependent tasks is vital. First, nonprescribed tasks, which have traditionally been performed by supervisors alone, must be made the responsibility of the work group; second, the power to negotiate interdependent performance agreements must be shared with those actually performing the key tasks. Unless these changes are made, interdependent parties will continue to refer performance problems to their superiors, who in turn will hand down temporary directions for conflict settlement.

Because power is a function of the dependence of one party on another, to the extent that power interferes with mutual cooperation it should be redistributed. This may call for a revision of task responsibilities, e.g., production operators can overcome their dependence on warehouse personnel by procuring their own resources. At a managerial level, this may call for a change in organizational structure. An example of such a change is provided by Lourenco and Glidewell (1975) in their analysis of shifts in the power of a major television network from the central office to its affiliates.

The following propositions sum up the relationship between power and task performance:

Proposition 14. The power an individual or group has to affect performance agreements is a function of the nature of the task(s) performed.

Proposition 15. To the extent that a power differential exists between parties to a performance agreement and that differential interferes with mutual cooperation on the task, organizational performance is hindered.

Proposition 16. Supervisors and professionals often hold more power than necessary to affect interdependent performance agreements. To the extent that the responsibility for negotiating performance agreements can be transferred to the parties involved, task performance becomes more dependable and organizational performance improves.

Proposition 17. To the extent that interdependent performance agreements can be made more prescribed, the influence of power differentials on performance contracts is reduced and task performance becomes more dependable.

Variables Affecting Key Interdependent Task Performance	*Conditions Facilitating Key Interdependent Task Performance*	*Conditions Impeding Key Interdependent Task Performance*
1. Effort—reward performance contracts	Perceived equity	Perceived inequity
2. Consistency of task performance by parties	Consistent	Inconsistent
3. Nature of task interrelationships	Prescribed	Nonprescribed
4. Nature of technology	Physical or certain: Defines interdependencies clearly	Nonphysical or uncertain: Discourages definition of interdependencies
5. Organizational structure	Rigid: Specifies interdependencies	Flexible: Interdependencies are not clear
6. Power differential between parties	Low: Facilitates equitable performance agreements	High: Impedes equitable performance agreements

Figure 1. Summary of Propositions

The elimination of supervisory roles or the abolishment of all organizational power differentials is not required. Roles and differentials will always exist either formally or informally, as they should. But the rational use of power to facilitate the negotiation of key task performance agreements for the benefit of the organization is necessary. The key propositions are summarized in Figure 1.

CONCLUSION

Power, structure, technology, and the nature of tasks interact to affect the performance of interdependent tasks that are crucial to an organization's success. A diagram of the theoretical relationships between these variables follows.

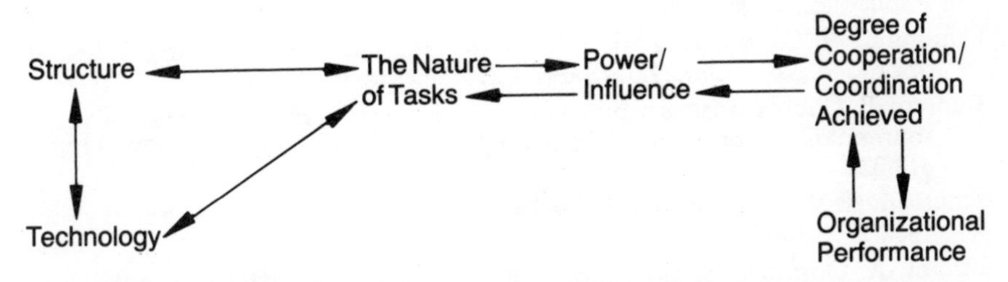

The relationships between the variables in this model are sequential; once given a structure and technology (which evolve almost simultaneously), certain tasks are prescribed and others are not. Through the use of power and influence, organization members make agreements to perform tasks that are not necessarily prescribed but are important to the organization. The degree of cooperation and coordination achieved directly affects the degree of organizational success.

If a change occurs internally or externally that makes new demands on the organization, the feedback effect reverses. Organizational performance suffers because certain tasks are not performed, which indicates that there is insufficient cooperation or coordination between the organization's internal systems. To correct this situation, supervisors are called on to use their power and influence to force the performance of certain tasks. If these tasks are not performed dependably on a discretionary basis, the nature of the tasks may have to be prescribed, and new employees may need to be hired to perform them. If the change is great and cannot be handled by the above methods, changes in the technology or the structure of the organization will take place; the organization will be remolded to fit better in its relevant environment and to perform its key tasks. In this way, the organization is dynamic, always adjusting to the needs of its internal and external environments. Future research must be done to examine the utility of this theory.

REFERENCES

Alderfer, C. P. Job enlargement and the organizational context. *Personnel Psychology*, 1969, *22*, 418-26.

Barnard, C. I. *The functions of the executive*. Cambridge, Mass.: Harvard University Press, 1938.

Bucklow, M. A new role for the work group. *Administrative Science Quarterly*, 1966, *11*, 59-78.

Burack, E. *Organizational analysis: Theory and applications*. Hinsdale, Ill.: Dryden Press, 1975.

Burns, T., & Stalker, G. *The management of innovation*. London: Tavistock Publications, 1961.

Cummings, T. G., & Srivastva, S. *The management of work: A sociotechnical approach*. Kent, Ohio: Kent State University Press, 1977.

Duncan, R. Characteristics of organizational environments and perceived environmental uncertainty. *Administrative Science Quarterly*, 1972, *17*(3), 313-328.

Emerson, R. M. Power-dependence relations. *American Sociological Review*, 1962, *27*, 31-40.

Etzioni, A. Organizational control structures. In J. March (Ed.), *Handbook of organizations*. Chicago: Rand McNally, 1965.

Galbraith, J. R. Environmental and technological determinants of organization design. In J. Lorsch and P. Lawrence (Eds.), *Studies in organizational design*. Homewood, Ill.: Richard D. Irwin, 1970.

Glidewell, J. C. *Choice points: Essays on the emotional problems of living with people*. Cambridge, Mass.: M.I.T. Press, 1970.

Katz, D., & Kahn, R. *The social psychology of organizations*. New York: John Wiley, 1966.

Krone, C. B. Open system redesign. In J. Adams (Eds.), *Theory and method in organization development: An evolutionary process*. Arlington, Va.: NTL Institute, 1974.

Lawler, E., Hackman, J., & Kaufman, S. Effects of job redesign: A field experiment. *Journal of Applied Social Psychology*, 1965, *49*, 24-33.

Lourenco, S., & Glidewell, J. A dialectical analysis of organizational conflict. *Administrative Science Quarterly*, 1975, *20*, 489-508.

March, J., & Simon, H. *Organizations*. New York: John Wiley, 1958.

Miller, J. G. The nature of living systems. *Behavioral Science*, 1971, *16*, 278-301.

Perrow, C. *Complex organizations*. Glenview, Ill.: Scott, Foresman, 1972.

Scott, W. G. Organization theory: An overview and appraisal. *Academy of Management Journal*, 1961, *4*, 7-26.

Sherwood, J. J., & Glidewell, J. C. Planned renegotiation: A norm-setting OD intervention. In W. W. Burke (Ed.), *Contemporary organization development: Conceptual approaches and interventions*. La Jolla, Calif.: NTL Learning Resources Corporation, 1972.

Simon, H. *Administrative behavior*. New York: Macmillan, 1975.

Srivastva, S., Salipante, P., Cummings, T., Notz, W., Bigelow, J., & Waters, J. *Job satisfaction and productivity*. Cleveland: Case Western Reserve University, 1975.

Thibaut, J. W., & Kelley, H. H. The social psychology of groups. New York: John Wiley, 1959.

Thompson, J. D. *Organizations in action*. New York: McGraw-Hill, 1967.

Walton, R. E. How to counter alienation in the plant. *Harvard Business Review*, November-December, 1972, pp. 70-81.

Weber, M. *From Max Weber, essays in sociology*. (Hans Gerth & C. Mills, Ed. and trans.) New York: Oxford University Press, 1947.

Weber, M. *The theory of social and economic organization*. (A. Henderson & T. Parsons, Ed. and trans.) New York: Oxford University Press, 1947.

Woodward, J. *Management and technology*. London: Her Majesty's Stationery Office, 1958.